An Introduction to District Heating and Cooling

Low carbon energy for buildings

Online at: https://doi.org/10.1088/978-0-7503-5286-4

IOP Series in Renewable and Sustainable Power

The IOP Series in Renewable and Sustainable Power aims to bring together topics relating to renewable energy, from generation to transmission, storage, integration, and use patterns, with a particular focus on systems-level and interdisciplinary discussions. It is intended to provide a state-of-the-art resource for all researchers involved in the power conversation.

Series Editor
Professor David Elliott
Open University, UK

About the Editor
David Elliott is emeritus Professor of Technology Policy at the Open University, where he developed courses and research on technological innovation, focusing on renewable energy policy. Since retirement, he has continued to write extensively on that topic, including a series of books for IOP Publishing and a weekly blog post for *Physics World* (physicsworld.com/author/david-elliott)

About the Series
Renewable and sustainable energy systems offer the potential for long-term solutions to the world's growing energy needs, operating at a broad array of scales and technology levels. The IOP Series in Renewable and Sustainable Power aims to bring together topics relating to renewable energy, from generation to transmission, storage, integration, and use patterns, with a particular focus on systems-level and interdisciplinary discussions. It is intended to provide a state-of-the-art resource for all researchers involved in the power conversation.

We welcome proposals in all areas of renewable energy including (but not limited to): wind power, wave power, tidal power, hydroelectric power, PV/solar power, geothermal power, bioenergy, heating, grid balancing and integration, energy storage, energy efficiency, carbon capture, fuel cells, power to gas, electric/green transport, and energy saving and efficiency.

Authors are encouraged to take advantage of electronic publication through the use of colour, animations, video, data files, and interactive elements, all of which provide opportunities to enhance the reader experience.

Do you have an idea for a book you'd like to explore?
We are currently commissioning for the series; if you are interested in writing or editing a book please contact Caroline Mitchell at caroline.mitchell@ioppublishing.org.

A full list of titles published in this series can be found here: https://iopscience.iop.org/bookListInfo/iop-series-in-renewable-and-sustainable-power#series.

An Introduction to District Heating and Cooling

Low carbon energy for buildings

Paul Woods, MA MSc CEng FIMechE FEI FCIBSE

IOP Publishing, Bristol, UK

ISBN 978-0-7503-5286-4 (ebook)
ISBN 978-0-7503-5284-0 (print)
ISBN 978-0-7503-5287-1 (myPrint)
ISBN 978-0-7503-5285-7 (mobi)

DOI 10.1088/978-0-7503-5286-4

Version: 20231201

IOP ebooks

British Library Cataloguing-in-Publication Data: A catalogue record for this book is available from the British Library.

Published by IOP Publishing, wholly owned by The Institute of Physics, London

IOP Publishing, No.2 The Distillery, Glassfields, Avon Street, Bristol, BS2 0GR, UK

US Office: IOP Publishing, Inc., 190 North Independence Mall West, Suite 601, Philadelphia, PA 19106, USA

To Kim, Owen and Jenny

Contents

Part II The Engineering of DH and DC systems

Preface

I began working in the field of District Heating (DH) in 1983, 40 years ago. I was very fortunate to join Orchard Partners and work with William Orchard and John Macadam. At that time there were hopes that there would be a national programme of investment in DH as part of an energy conservation strategy. Since then, there have been a number of false dawns for the industry but in 2013 the Government published its first national heat strategy which identified heat networks as part of the long-term solution to achieve net zero emissions by 2050. Ten years later the Government's energy strategy described heat networks as 'vital to making net zero a reality'. A number of studies have estimated the economic potential for heat networks to be between 18% and 43% of the total heat demand. Whichever figure proves to be correct this still represents a huge increase in capacity from the current market share of around 3%. To deliver this capacity will require many more skilled people to enter the industry, whether that is engineers, technicians, project managers, lawyers, financiers, commercial advisers or marketing specialists. The primary purpose of this book is to help new entrants, and especially engineers, gain some essential knowledge quickly so that they can contribute to realising the national potential for DH.

Structure of the book

The book is in three parts:

Part I provides a general introduction to district heating and cooling (DH/C) with chapter 1 providing some historical context and chapter 2 describing the physical components that make up a typical DH/C scheme. It is intended that this will be of interest to a non-technical readership as well as providing a useful background for engineers.

Part II discusses the engineering of DH/C beginning in chapter 3 with heating and cooling systems within buildings and how existing systems may need to be modified to suit DH/C. This is followed by chapter 4 which shows how the interface between the DH/C networks and the building should be designed to deliver an efficient system. Chapter 5 discusses the piped networks themselves, the technologies used and the main design aspects including: pumping and pressures, heat losses, optimal pipe sizing, stress analysis and quality control. Chapter 6 reviews the wide range of heat sources that are now available for heat networks, starting with gas-fired CHP which has been the commonest heat source for some time, followed by heat pumps, heat recovery and renewable energy sources. Chapter 7 focuses on the strategic design decision of the selection of operating temperatures and puts forward proposed temperatures that would be suitable for a heat pump based system supplying existing buildings. This part of the book concludes with chapter 8 which discusses some operational aspects including water treatment and heat metering both of which are critical to a successful scheme. In Part II I have sought to explain how fundamental engineering concepts can be applied to the challenges of DH/C to form a bridge between theoretical knowledge and the practical problems faced by engineers delivering projects. I have included 11 fundamental equations that an engineer will need to be familiar with, however, these have been kept to a minimum. This part of the book may be useful in supporting taught courses in DH/C or more generally in mechanical engineering degree courses.

Part III has a broader scope than the engineering of DH/C. Chapter 9 sets out how to make a case for DH/C beginning with the well-known energy trilemma of environment, cost and security of supply. Some other potential advantages of DH/C to potential customers are also explored. Chapter 10 looks at a wide range of new opportunities that DH/C could profit from in the future. The role that DH/C could have in helping to manage the challenge of increasing amounts of variable renewable electricity generation is assessed using a simplified model of future generation and demand patterns. This shows that there is a case for increased amounts of thermal energy storage and that gas-fired CHP may still have a valid role into the 2030s. The chapter also considers other long-term options that the Government has identified in the 2023 energy strategy 'Powering up Britain' including: hydrogen, small modular reactors and carbon capture usage and storage and how DH/C could be linked to these initiatives. With climate change leading to increasing temperatures, a section examines the ways in which DH can be linked with DC to improve efficiency, including the use of ambient temperature networks. The new Government policy on zoning of cities to promote rapid growth of DH is discussed. The chapter concludes with some ideas on how the costs of DH can be reduced so that customers will be better served and market share can be increased.

Chapter 11 contains a summary of some of the themes of the book and why I think the case for DH/C remains as strong as ever.

District cooling

Although the title of the book is *District Heating and Cooling*, inevitably there is more of an emphasis on heating than cooling. This is partly because much of the technology is either suitable for both heating and cooling or similar to both. However, I am aware that there is much more detail on DC that could have been included and I would refer the reader to two valuable sources of further information on DC:

ASHRAE District Cooling Guide (2nd edn) and Owner's Guide for Buildings Served by District Cooling, ASHRAE, 2019

Sustainable District Cooling Guidelines, IEA, 2020 https://www.iea-dhc.org/the-research/annexes/annex-xii/annex-xii-project-05.

Website links

Where possible I have provided hyperlinks to references in the Bibliography, especially for Government reports which can be accessed easily. I have included the link to the main site rather than the actual document referenced as often the site contains other useful guidance and references.

In some parts of the book, especially chapter 6, I have included some hyperlinks in the text. These are links to websites about projects, organisations or companies of interest rather than to specific documents. I have selected sites that provide the most useful information about a particular project—not necessarily the website of the owner of the scheme.

Devolved governments

I am aware that in dealing with Government policy this has been based on policies applicable to England. There are differences in the ways in which DH/C is being supported and developed by the devolved governments. My lack of knowledge of these differences and the limited reference to these does not imply that they are not important. In fact, in many areas the devolved Governments are leading the way for the UK. The information is as up to date as possible at the time of writing but this aspect of the book will obviously become out of date as policies develop.

Further reading

I have made many references to the CIBSE/ADE Code of Practice CP1.2 which lists requirements to be followed and compliance with CP1.2 is often a condition of a grant award or a planning application. It is hoped that this book will provide additional background so that the reasoning behind the requirements of CP1.2 can be better understood (see bibliography for chapter 3 reference [9]).

The very comprehensive textbook *District Heating and Cooling* by Svend Frederiksen and Sven Werner has been particularly valuable to me in writing this book and it is highly recommended for further reading, especially as it provides a Scandinavian perspective (see bibliography for chapter 1 reference [6]).

In addition to the sources in the Bibliographies, which are provided at the end of each chapter, the following websites are very useful for general guidance and information on the industry and many have extensive resources available to download.

Useful websites

The Association for Decentralised Energy https://www.theade.co.uk/

Euro Heat and Power https://www.euroheat.org

Heat Networks Delivery Unit https://www.gov.uk/guidance/heat-networks-delivery-unit

Heat Network Support Unit Scotland https://www.heatnetworksupport.scot/

International Energy Agency, District Heating and Cooling https://www.iea-dhc.org/home

UK District Energy Association: https://ukdea.org.uk/.

Acknowledgments

I am very grateful to the Institute of Physics for the opportunity to write this book and the encouragement of the Series Editor David Elliott. Thank you to Caroline Mitchell and Isabelle Defillion for all their help and advice which made the whole process easier than I had envisaged.

I have only been able to write this book because of the knowledge I have gained through working with excellent colleagues over the last 40 years. These include: at Orchard Partners, William Orchard and John Macadam, at Steensen Varming Mulcahy, Dusan Markovic, Michael Carver and Bryan Franklin, at Parsons Brinckerhoff, Dominic Bowers and Ian Burdon, at AECOM, Richard John and Ant Wilson, at Islington Council, Lucy Padfield, at Engie, David Culver and Ben Watts, and at Sustainable Energy, Gabriel Gallagher, Lee Evans and Chrissy Woodman. I would also like to pay tribute to the late Jens Overgaard of Ramboll from whom I learnt much. It has been a pleasure working with many other colleagues in the industry, whether as co-workers, in client organisations or with contractors and suppliers, too numerous to mention individually, although the dedication of Michael King and Phil Jones have been inspirational.

I am very grateful to all who kindly reviewed draft chapters of the book and provided valuable comments. The book is greatly improved as a result of your work and I enjoyed catching up with old friends in the process. So, my special thanks to: Ian Allan, Professor Mark Barrett, David Burke, Peter Concannon, Martin Crane, Annabel Harford, Rufus Ford, Dr Gabriel Gallagher, Bruce Geldard, Gareth Jones, Charlotte Large, John Macadam, Dr Anthony Riddle, Oliver Riley, Andy Simms, Phillippe Terrine, Stephen Ward, Mark Whettall, David Whitfield, Dr Robin Wiltshire and Owen Woods. Of course, the responsibility for the final text lies entirely with myself.

I would like to thank my children, Owen and Jenny and my wider family for their encouragement. Finally, thank you to my wife Kim for her expert advice (as the author of three books herself) and kind support for this book as for everything else in my life; I could not have completed this without you.

Legal disclaimer

Author biography

Paul Woods

Paul Woods graduated in Engineering from Cambridge University and, after initial training in the nuclear power industry, started working in District Heating (DH) for Orchard Partners in 1983. He took a part-time MSc at Cranfield University in energy and the environment. He continued to work in DH, District Cooling (DC) and the broader field of energy and buildings, taking lead roles in the design of the building services of the New British Library (with Steensen Varming Mulcahy) and the DH/C systems for Heathrow Terminal 5 (with Parsons Brinckerhoff). He was technical advisor to the UK Government for the implementation of the Energy Performance of Buildings Directive (with AECOM). He led two international research projects for the IEA District Heating and Cooling programme and was Chief Technology Officer for two research projects in DH for the Energy Technologies Institute. In 2009, he co-authored the report on the Potential and Costs of DH (known as the Poyry report) which informed the Government's heat strategy. He was seconded to Islington Council to develop the Bunhill DH project. International experience has included working on DH projects in Turkey, the Republic of Ireland, Australia, South Korea, Romania, Russia and DC projects in Malaysia and India. Paul represented the UK on the European CEN standards committee for DH pipe systems and chaired the BSI committee RHE9/3. He was editor of CIBSE's Application Manual for CHP (AM12) and author of the first edition of the CIBSE/ADE Code of Practice for Heat Networks CP1 and co-author of the second edition. At Engie, Paul was appointed Concessions Director and was responsible for the strategic development of their six UK district heating and cooling schemes, including at the Queen Elizabeth Olympic Park (QEOP), later taking on an operational role at QEOP. Since 2017, after moving to Brecon, Paul has been working as an independent consultant on a number of DH and DC projects including in Cardiff, Bristol, South Dublin, Solihull and Bridgend and is an Associate with Sustainable Energy Ltd.

Symbols

A	cross-sectional area of pipe (m^2)
A	area of heat exchanger (m^2)
c	specific heat capacity (kJ kg^{-1} K)
d_i	pipe internal diameter (mm)
d_o	pipe external diameter (mm)
e_e	electricity emissions factor
e_g	gas emissions factor
e_{grid}	grid emission factor
E	Young's modulus (kN mm^{-2})
E_w	work efficiency
E	electricity output
F	friction on pipe in ground (N m^{-1})
F_{chp}	fuel input
g	gravitational acceleration (m s^{-2})
H	heat output
h	height (m)
k	thermal conductivity (W $m^{-1}K^{-1}$)
k	roughness coefficient (m)
$\dot{m}$	mass flow rate (kg s^{-1})
P	power (kW)
p	pressure (N m^{-2})
Δp	pressure difference (N m^{-2})
Q	heat energy flow rate (kW)
T	temperature (K)
ΔT	temperature difference (°C)
t_w	pipe wall thickness (mm)
U	heat transfer coefficient (W m^{-2}K)
v	velocity (m s^{-1})
V	volume flow rate (m^3 s^{-1})
z	z-factor for steam turbine extraction
α	temperature coefficient of linear expansion (K^{-1})
η_b	boiler efficiency (%)
η_{grid}	grid efficiency (%)
η_e	CHP electrical efficiency (%)
η_h	CHP heat efficiency (%)
λ	friction factor
μ	dynamic viscosity (Ns m^{-2})
ρ	density (kg m^{-3})
σ	stress (N mm^{-2})

Glossary of acronyms

ADE	The Association for Decentralised Energy
AGR	advanced gas-cooled ractor
AHU	air-handling unit
ATES	Aquifer Thermal Energy Storage
ASHP	air source heat pump
ASHRAE	American Society of Heating Refrigeration and Air Conditioning Engineers
BEIS	Department for Business, Energy and Industrial Strategy
BEMS	building energy management system
BESA	Building Engineering Services Association
CCGT	combined cycle gas turbine
CCUS	carbon capture usage and storage
CEN	European Committee for Standardisation
CHP	combined heat and power
CIBSE	Chartered Institution of Building Services Engineers
CoP	coefficient of performance
DC	district cooling
DECC	Department of Energy and Climate Change
Defra	Department of Environment, Food and Rural Affairs
DESNZ	Department of Energy Security and Net Zero
DH	district heating
DH/C	district heating and cooling
DHW	domestic hot water
DPCV	differential pressure control valve
DUKES	Digest of UK Energy Statistics
ETI	Energy Technologies Institute
GB	Great Britain, i.e. the UK excluding Northern Ireland
GHNF	Green Heat Networks Fund
GSHP	ground source heat pump
GWP	global warming potential
HNDU	Heat Networks Delivery Unit
HIU	Heat Interface Unit
HNIP	Heat Networks Investment Project
HV	high voltage 11 kV
IRR	Internal Rate of Return
LA	local authority
LMTD	log mean temperature difference
LNG	liquefied natural gas
LV	low voltage 400 V
mtoe	million tonnes of oil equivalent
NPC	Net Present Cost
NRSWA	New Road and Street Works Act
PICV	pressure independent control valve
OCGT	open cycle gas turbine
OPEC	Organisation of Petroleum Exporting Countries
SIGE	spark-ignition gas-engine

SMR	small modular reactors
UNFCCC	United Nations Framework Convention on Climate Change
WRAS	Water Regulations Approval Scheme
WSHP	water source heat pump

Part I

Introduction to district heating and cooling

IOP Publishing

An Introduction to District Heating and Cooling
Low carbon energy for buildings
Paul Woods, MA MSc CEng FIMechE FEI FCIBSE

Chapter 1

Introduction

1.1 What is district heating and cooling?

Most buildings in the northern hemisphere need a form of heating to maintain adequate levels of comfort. From the earliest times, wood was used as fuel for open fires to keep warm, being replaced by charcoal and then by coal. Larger buildings began using 'central' heating in the mid-19th century where instead of fires in each room a single boiler would distribute heat using hot water. Initially these boilers used coal and later oil or natural gas. District heating (DH), also known as heat networks, takes this process a step further by centralising the production of heat and distributing it to a number of buildings using insulated pipe networks.

District heating is the distribution of heat to a number of buildings from one or more heat production plants

Within this broad definition there is a wide range of scale:

- Small housing developments
- Larger developments of multiple flats in one or more blocks (sometimes referred to as communal or group heating systems)
- Campus based heat networks often used for university or hospital sites and defence establishments
- Small district schemes connecting a number of buildings both commercial, institutional and residential
- Large district schemes extending across significant areas of cities or city-wide

Examples of large DH schemes include the cities of: Moscow, Copenhagen, Helsinki and Stockholm, which have extensive heat networks supplying most of their city's buildings.

In the UK, smaller city centre DH schemes can be found in many cities including: London, Sheffield, Nottingham, Leeds, Manchester, Aberdeen, Birmingham, Coventry, Leicester, Bristol, Southampton and Cardiff. Although most of these

doi:10.1088/978-0-7503-5286-4ch1

schemes currently supply a relatively small share of the city's heat demand there is obviously potential for significant expansion.

Smaller towns with DH networks include Lerwick, Bridgend and the new town being built at Cranbrook, Exeter. Campus DH schemes include the universities of Warwick, Loughborough, Liverpool, Edinburgh and Reading. The Slough industrial estate is an example of a major industrial DH scheme.

Since 2004, the energy hierarchy in the London Plan [1] has effectively required most large developments in the capital to be heated by a communal hot water based system so that they can reduce their energy use and be connected to a wider DH system in the future.

Currently DH provides about 3% of the UK heat demand [2].

Although the earliest district heating schemes used steam as the heat distribution medium, for example those in Copenhagen and New York, it is now universally preferred to use hot water.

District cooling (DC) is a similar concept to DH with cooling energy (sometimes termed coolth) being distributed to a number of buildings in the form of chilled water from a centralised chiller plant. There are many DC systems in the Middle East, Japan and China; in Europe there are projects in Paris and Stockholm, and, in the USA, there are examples in Minneapolis-St Paul and Chicago. In the UK there are DC networks at Citigen in the City of London, the Queen Elizabeth Olympic Park and Terminal 5 Heathrow Airport.

DH/C is a method for transporting energy but is not in itself an energy source. So, when discussing the role of DH/C in energy strategy it is always important to consider the energy source(s) that supply the DH/C systems.

Figure 1.1 is a diagram produced by Ramboll [3] to explain the concept of DH/C. A general introduction to DH/C and its future role in combating climate change is

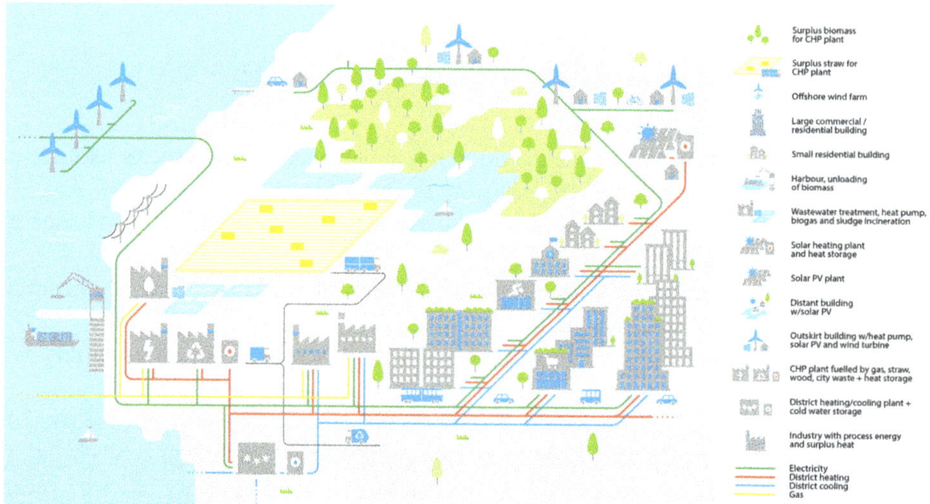

Figure 1.1. A representation of the concept of district heating. Reproduced with permission from Ramboll.

presented in a video presentation produced by the European DH/C trade body, Euroheat and Power [4].

1.2 Why use DH/C?

1.2.1 Energy efficiency

Since the first oil crisis of the 1970s, energy security has been a major concern. In the UK we were shielded by having indigenous energy resources of coal as well as oil and gas from the North Sea. Other countries were more dependent on imported fuels and from the 1970s they started to explore opportunities to improve the efficiency of fuel use.

The conversion of fuel to electricity in power stations has a relatively low efficiency at 40%–50%. This is a function of the temperature limitations of materials and thermodynamic fundamentals. Around 30%–40% of the original energy in the fuel is discharged as low temperature (ca. 30 °C) heat via cooling towers or into the sea. However, it is straightforward to modify the power stations to produce heat at a more useful temperature, as well as electricity, and this has enabled significant gains in energy efficiency. This concept is known as combined heat and power (CHP). Many CHP plants, initiated in the 1960s and 1970s, are used to supply city-wide DH schemes in the USSR, Eastern Europe and Scandinavia.

When heat is produced at a higher temperature from a power station there is some reduction in electricity generation which will have to be replaced by additional electricity production elsewhere in the system. However, the additional fuel used for this is much less than the fuel that would have been consumed in boilers in individual buildings. The amount of additional electricity required can be defined by the '*z*-factor':

$$z = \text{useful heat/reduction in electricity}$$

Heat from a CHP plant can be compared with heat from other heat sources such as a boiler or a heat pump in three ways:

A. Comparison of efficiencies which are commonly quoted as for a boiler, i.e., heat output divided by fuel input
B. Comparison of fuel used to produce a unit of heat—this is the inverse of the efficiency
C. CO_2 content of heat from each source

In the comparison below we have used method B; for the other comparisons see section 6.4.

Boiler heat
Suppose a boiler is 80% efficient, this means that 1.25 units of fuel are required per unit of heat.

Heat from a power station
The *z*-factor is the ratio of the useful heat produced to the reduction in electricity that occurs. For a typical *z*-factor of 8 then for one unit of heat supplied, 0.125 units of electricity is 'lost' and needs to be produced elsewhere. Assuming an electricity generation efficiency of 50%, 0.25 units of fuel would be needed for this electricity. Hence, 0.25 units of fuel is needed per unit of heat produced. This heat will need to be transported through a DH system and as a 10% heat loss is typical this results in 0.275 units of fuel per unit of heat delivered.

Heat pump
The performance of a heat pump is defined by its coefficient of performance (CoP) which is the ratio of useful heat produced divided by electricity used. With a CoP of 3, 0.33 units of electricity is needed per unit of heat. With electricity produced at 50% efficiency and 45% after allowing for grid losses, this means 0.74 units of fuel is used per unit of heat.

The above results are shown in figure 1.2.

It can be seen that heat extracted from power stations (large-scale CHP) uses about *a fifth of the fuel needed compared to a boiler.* This was the opportunity that was implemented from about 1970 by many cities in the USSR, Eastern Europe, Denmark, Sweden and Finland, but not by the UK.

The use of CHP with DH has remained one of the major drivers for DH. As a result of the energy efficiency improvement with CHP, CO_2 emissions are also lower

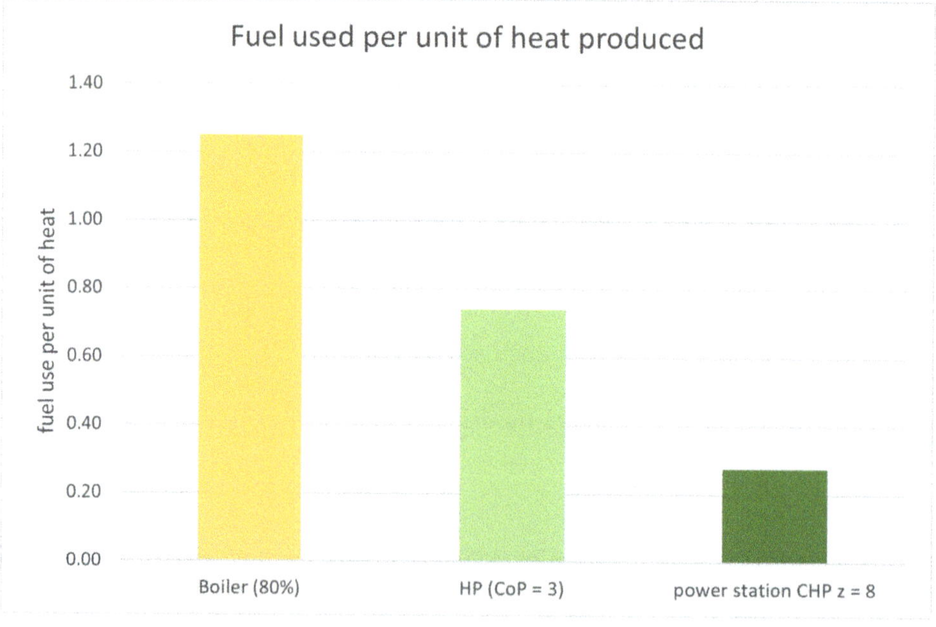

Figure 1.2. Comparison of fuel needed per unit of heat supplied for boilers, heat pump and power station CHP with DH.

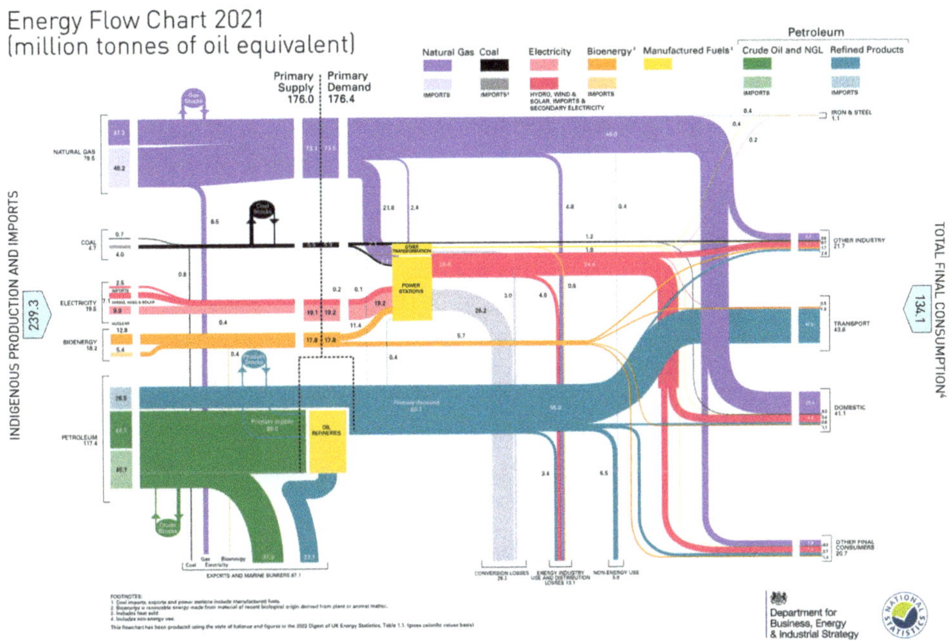

Figure 1.3. Sankey Diagram for the UK's energy flows. Reproduced from [5]. Contains public sector information licensed under the Open Government Licence v3.0.

and this became another important policy aim in most countries following the UN Conference on Environment and Development held in Rio de Janeiro in 1992.

The amount of heat lost from all power stations in the UK can be seen from the Sankey diagram which is published each year with the 2021 version presented as figure 1.3 [5]. The diagram shows all energy flows with the width of the bars related to the amount of energy.

From figure 1.3, the amount of heat rejected from power stations was similar in scale to the total energy in the gas supplied to all domestic buildings (26.2 mtoe compared to 27.4 mtoe).

Smaller scale CHP also results in significant energy savings and CO_2 reductions. As the smaller CHP plant is located closer to the buildings supplied, the losses associated with electricity and heat distribution are lower. For the assumptions on efficiency given in figure 1.4, 100 units of input fuel for the CHP plant displaces 133 units of fuel if the heat and electricity are produced separately, a saving of 25% in primary energy use and associated CO_2 emissions.

1.2.2 Advantages of DH/C

In addition to the use of CHP there are many other reasons why DH/C may be advantageous depending on local circumstances:

- Some heat or coolth sources may be available at low cost and low CO_2 content but only in specific locations, e.g., heat recovery from an industrial

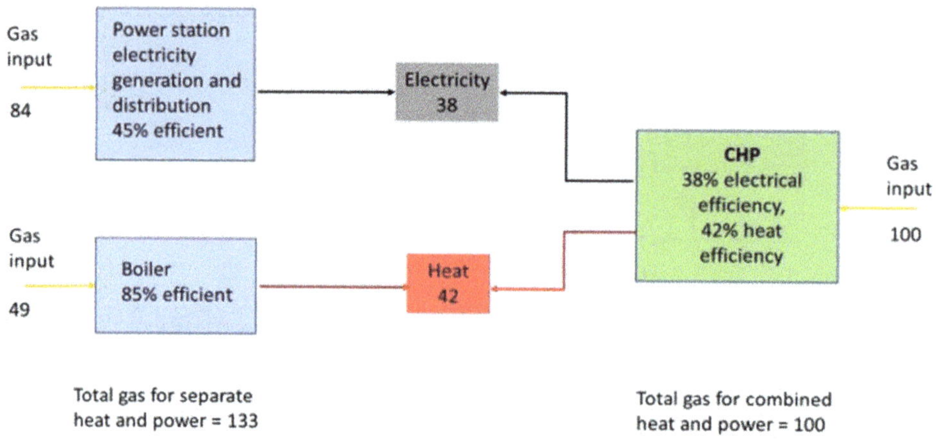

Figure 1.4. Savings associated with small-scale gas-fired CHP.

process, a major datacentre, or for DC, cool water from snowmelt at depth in a river.

- Centralised heat/coolth production can benefit from economies of scale in terms of both construction cost and operational efficiency, e.g., heat pumps are cheaper and more efficient at a larger scale than for individual buildings.
- Alternative fuels may be available which can only be utilised in large, specialised facilities not normally at a building level (e.g., heavy fuel oil, waste, woodchips).
- A larger scale energy plant supplying DH/C will be able to purchase fuel and electricity more cheaply
- Better control of the pollution associated with heat/coolth production can be achieved in a centralised plant rather than at a building level.
- Multiple heat sources of different types can be used to provide an optimal solution which tends to be impractical or expensive at a building level.
- Thermal storage can be introduced at a larger scale than is normally possible at a building level; this can result in lower operational costs.
- Large centralised thermal storage can provide flexibility benefits to the national electricity system enabling greater use of variable renewable electricity generation.
- Inter-seasonal heat storage can be incorporated especially in conjunction with large-scale solar thermal heat sources.
- The provision of both heating and cooling from an integrated system can bring additional energy efficiency benefits.
- DC systems may be able to use evaporative cooling towers which improves efficiency and may not be feasible at a building level.
- DC or heat pump based DH systems can use a wider range of refrigerants with lower global warming impacts, as the potential safety issues associated with some refrigerants can be more easily managed in a large central facility.

As a result of many of the above benefits, CO_2 and other greenhouse gases emitted in the process of producing heat and coolth can be reduced significantly.

1.2.3 Potential drawbacks

The above benefits need to result in a sufficient reduction in cost compared to individual building systems to justify the investment in the heating/cooling distribution infrastructure.

Development of DH/C requires a large proportion of building owners in a given area, if not all, to commit to taking heat otherwise the scheme would be uneconomic. DH/C has developed fastest in countries where energy supply is centrally planned rather than solely market driven. In these circumstances, DH/C becomes a natural monopoly and a suitable regulatory framework is needed that protects customers from high prices or poor service standards.

1.3 Development of district heating

1.3.1 International development

The origins of modern DH began in the USA in the 1880s when a number of steam systems were built including the Manhattan scheme which is still in operation delivering 6900 GWh of heat in 2009 [6] (1 GWh is about the heat requirement of 100 houses in the UK). The steam supply is also used by absorption chillers for air-conditioning. DH using steam remained a popular solution in the USA but mainly for campus applications such as: universities, hospitals and military bases. One exception to this is the hot water DH scheme in Minneapolis-St Paul developed from 1979 and inspired by Scandinavian experience which in 2021 supplied 330 GWh across the city [7].

In Europe, the first major schemes were built in Germany in the 1920s in response to the high price of coal after the first world war, including those in Hamburg, Kiel, Leipzig and Berlin. Projects in Copenhagen, Utrecht and Paris followed. The Reykjavik DH scheme using geothermal heat began in 1930 and now supplies all the buildings in the city.

However, most DH developments in North-Western Europe took place from the 1970s when OPEC increased the price of oil threefold. For countries that depended heavily on oil for heating this created an urgent need to look for other solutions. At the time, Denmark, Sweden and Finland did not have access to natural gas and imported all the coal needed for power generation; CHP supplying DH offered a practical and efficient alternative to oil for heating.

DH developed rapidly within Russia and Eastern Europe, with very large schemes in Moscow, Bucharest and East Berlin, aided by the centrally planned nature of these economies. Most of these schemes utilised heat extracted from major power stations. The Moscow DH system expanded from a peak supply of 5060 MW in 1960 to 27 000 MW in 1981 [8] and is considered the largest in the world, with 120 000 GWh of heat now supplied annually. About 75% of housing in Russia is connected to DH [9]. The DH share of the residential heat market is about 65% in Denmark, 59% in Sweden, 38% in Finland and 14% in Germany [10].

The cost of heat distribution was initially a barrier, but this was addressed with the advent of pre-insulated pipe systems which were introduced in the 1970s. These enabled pipes to be buried directly in the ground instead of using expensive concrete ducts. Some of the early products proved unsuccessful and this led to failures and a loss of confidence especially in the UK. However, other countries persisted with the technology and improvements were made leading to the establishment of European standards beginning with EN253 (see section 5.6).

China has used DH extensively, historically based on large coal-fired boiler plants rather than CHP, although this is changing and there is now greater use of natural gas. Growth in recent decades has been very rapid. The Beijing area has 5000 DH systems supplying a total of 101 000 GWh in 2009. One of the main drivers for DH in China is to displace small coal-fired boilers which cause significant air pollution and have associated health impacts [9].

There are also large DH schemes in the Far East including in Seoul, Republic of South Korea where heat is supplied from a large combined cycle gas turbine (CCGT) power station.

For further information on the development of international DH schemes see [6, 9].

1.3.2 Development of district heating in the UK

The 1970s saw the development of the North Sea as a source of natural gas which not only displaced town gas but rapidly increased its share of the heating market in the UK.

In the 1960s and 1970s many new power stations were built. These were large-scale and if coal-fired were often located near the coalfields of Nottinghamshire and Yorkshire, e.g., West Burton A, Ferrybridge C or if oil-fired at ports, e.g., Fawley, Isle of Grain and so most were too remote from major population centres for DH to be considered. The nuclear power stations were also located on remote sites for safety reasons.

Two exceptions were the Pimlico scheme which supplied the 3000 dwellings in the Churchill Gardens estate by taking heat from Battersea power station from 1950 and the energy from waste plant which supplied the St Ann's estate in Nottingham, which was developed from 1974.

Most smaller DH developments were boiler-only schemes. Many of the postwar housing estates built in the 1960s and early 1970s used DH with low-cost heavy fuel oil (HFO) in preference to gas as there was a significant cost benefit and to avoid the use of the more polluting coal. As the price of gas fell with the advent of natural gas and oil prices rose this benefit was eroded which put pressure on these DH schemes; many were removed and the estates converted to gas boilers in individual dwellings.

Another major DH project constructed in the 1960s was the Whitehall scheme which supplies most of the major Government buildings (including Downing St) from a boiler plant located in the Ministry of Defence building. This plant was subsequently redeveloped to include a gas turbine CHP plant.

Box 1.1. File:Churchill Gardens Estate, Churchill Gardens Road (geograph 2678683).jpg—Wikimedia Commons

A representation of the concept of district heating. This Churchill Gardens Estate, Churchill Gardens Road (geograph 2678683).jpg image has been obtained by the author(s) from the Wikimedia website where it was made available under a CC BY-SA 2.0 licence. It is included within this book on that basis. It is attributed to Stephen Richards.

The Pimlico scheme has an interesting history which demonstrates how heat supply plant can be changed over time without the customers being significantly affected. The scheme started operating in 1950 using heat taken from dedicated CHP steam turbines in Battersea power station on the other side of the Thames. After the power station was closed, coal-fired boilers were installed adjacent to the power station site to maintain the heat supply. These were later converted to gas and later still the link under the Thames was abandoned and new gas boilers installed on the north side of the river. Subsequently, gas-fired CHP units were added to the scheme.

A major national study known as the Marshall Report (the work was led by Sir Walter Marshall, later Lord Marshall of Goring) concluded in 1979 that there was a case for developing DH in the UK and recommended the formation of a National Heat Board. It also recommended that the uncertainties on cost and market penetration would best be resolved by developing a scheme in one city. This led to further analysis in 1983 to determine which city would be preferred as the 'Lead City'. A small amount of Government funding was subsequently allocated to three cities: Leicester, Newcastle and Belfast but these projects did not proceed. However, three other schemes did commence construction:

- The Southampton scheme (1986) that arose as a result of the adoption of a research well for deep geothermal heating, that had been financed by central government.
- Sheffield Heat and Power (1987) based on an existing energy from waste plant.
- The Citigen scheme in the City of London (which includes a district cooling network) (1993) using two dual-fuel (oil/gas) CHP engines.

A major publicly funded demonstration city-wide scheme was never started. Meanwhile, there was a policy to privatise the major energy companies starting with British Gas in 1986 and followed by the break-up of the Central Electricity Generating Board in 1991. The supply of natural gas from the North Sea expanded and became a low-cost heating fuel across most of the UK with 86% of heating now provided by gas boilers. As a result of these developments, the incentive and impetus for a major, nationally planned, DH strategy for heat supply was lacking.

After the Rio summit in 1992, the issue of climate change began to drive policy initiatives and the value of CHP in industry and DH became better recognised with the Government setting a target for 10 GW of CHP capacity (of which about 6 GW was realised). Most of the CHP capacity taken forward in the 1990s was to supply industrial sites rather than for DH.

In the 1990s, the privatisation of the electricity industry and the removal of the constraints on the use of gas for power generation led to a number of gas-fired power stations being built—the so-called 'dash for gas'. Some of these were large and located remotely from city centres, e.g., Little Barford, 698 MW (1996), however some were near cities and could have been used for DH if suitable planning controls had been introduced for example: the Barking Reach CCGT, 1000 MW (1995 demolished in 2018), Peterborough CCGT, 360 MW (1993) and Seabank 1 and 2 CCGT (near Bristol), 1140 MW (2000/1). In this period an opportunity was lost to use thermal power stations to supply DH which would have helped to conserve our North Sea gas reserves for many more years.

In 1993, the energy from waste plant South East London CHP (SELCHP) was built and designed as a CHP plant from the outset with the potential to supply heat to the high density housing estates in the local area, many of which were already heated from boiler-only DH schemes. The SELCHP plant was financed with Government support in the form of a premium electricity price (the non-fossil fuel obligation). This meant that heat extraction for DH would be uneconomic without a compensating payment—which was not forthcoming. This lost opportunity was only ended in 2014 when SELCHP started supplying heat for the first time to 2500 dwellings and now has plans to expand the network further.

Subsequently, DH projects in the UK have depended on a number of Government capital grant programmes which have resulted in important but smaller-scale projects. The Community Energy Programme ran from 2001 to 2007 and supported a number of projects including: Birmingham, Aberdeen, Edinburgh University, Woking and Pimlico in Westminster. The growth in the industry was supported by training courses and guidance published by the Energy Efficiency

Office and through international initiatives such as the IEA District Heating and Cooling Programme.

In Scotland, projects in Lerwick (1991) using energy from waste and Aberdeen (2002) using gas-engine CHP were constructed, with the Aberdeen networks expanding significantly since. Edinburgh University also expanded its DH schemes and added gas-fired CHP units.

From 2003, local authorities began setting planning requirements in relation to energy that went beyond the Building Regulations requirements. The London Borough of Merton started the process by requiring 10% of energy needs to be met by renewable energy systems. Larger developments in London need planning permission from the GLA, and the London Plan in 2004 [1] required these new developments to have centralised heating systems with CHP plant and/or other renewable energy sources. Part of the aim of this policy was to ensure that these new buildings could be easily connected to a wider DH system in the future. This was in contrast to the electric heating systems that had become the norm and which would be incompatible with DH.

From 2006, an area of Stratford in East London was developed initially for the 2012 Olympic Games venues and athletes' village but subsequently has been further developed by the London Legacy Development Corporation as a major commercial, residential, retail and educational district. An early strategic decision was taken that all buildings would be required to connect to the DH and DC networks. The DH/C system there currently supplies around 110 GWh of heat p.a. and 50 GWh of cooling (see section 6.7, box 6.5).

In 2009, The Low Carbon Infrastructure Fund was set up by the newly formed Department of Energy and Climate Change and administered by the Homes and Communities Agency. This provided £26 m of grant funding to support 16 projects including: Newcastle, Southampton, Exeter, Milton Keynes and Hackney. Following the completion of this programme a guidance document was produced [11].

In Islington, the Bunhill scheme was developed from 2012 with a grant from the GLA which linked together a number of existing housing estates already served by DH to be supplied from a 1.6 MW gas-fired CHP plant. This scheme was extended from 2016 in a very innovative project using a heat pump with its source of heat being air extracted from a local underground railway ventilation shaft [12]

In 2009, Poyry and Faber Maunsell were commissioned by the newly formed Department of Energy and Climate Change to review the opportunities for DH and their report entitled 'The potential and costs of district heating networks' was produced [13]. This work led to the first Government papers on heat strategy, culminating in 'The Future of Heating: Meeting the Challenge' published in 2013 [14]. For the first time, a national heat strategy was produced that recognised the role of DH as an important part of the strategy to achieve carbon targets in 2050. Figure 1.5 (redrawn from Figure 7 in the strategy) shows possible trajectories for both the growth of DH and building scale heat pumps.

The Heat Networks Delivery Unit (HNDU) [15] was formed in 2013 to support and promote the development of heat networks in England and Wales. In the November 2015 Spending Review the Government announced a new programme of

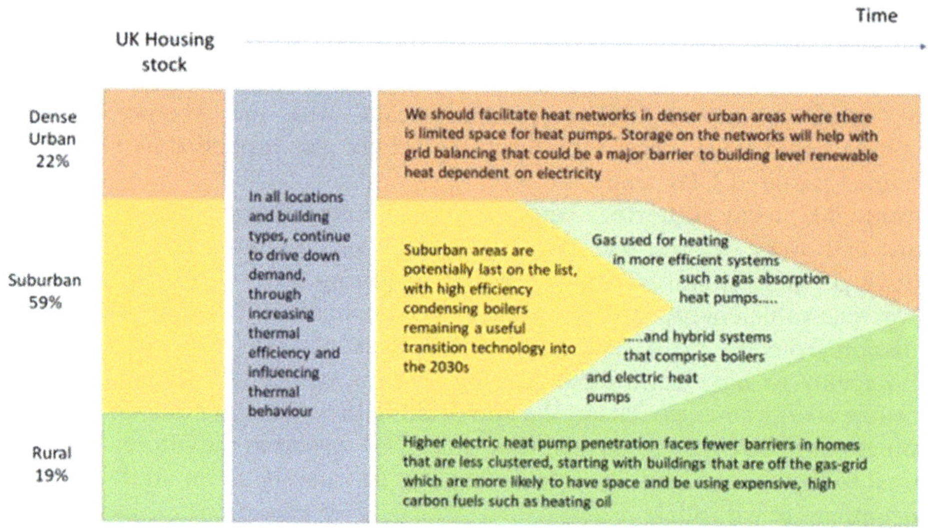

Figure 1.5. Strategic framework for low carbon heat in buildings over time.

grant support for heat networks of up to £320 m. This has funded feasibility studies, detailed project development work and provided capital grants, the latter being delivered from 2018 through the Heat Networks Investment Project (HNIP). Capital grants for construction have been awarded to a number of projects including those in: Bristol, Leeds, Haringey, Enfield, Southwark, Solihull, Gateshead, Barking, Manchester and Cardiff. A follow-on grant programme, the Green Heat Networks Fund (GHNF) was set up in 2021 with £288 m of funds allocated. The Heat Network Partnership for Scotland was set up in 2013 [16] to provide grants towards feasibility studies, to co-ordinate support and disseminate guidance.

For further information on the development of DH in the UK see section 1.1.2 in [9].

1.3.3 Future potential market share for DH

A number of studies have estimated the potential for DH in the UK.

The Energy Technologies Institute (ETI) macro-Distributed-Energy research work concluded in 2012 that 43% of the national heat demand would be economic to connect to DH [17].

In 2015, a report commissioned from Ricardo by the Department of Energy and Climate Change (DECC) [18] showed that the *technical* potential for DH was 59% of the total heat demand of the country and 81% of the supply of heat to buildings. How much of this potential could be realised depended strongly on the cost of financing, the cost of carbon and the costs of the alternative low carbon solutions.

An assessment of the potential for DH in the UK was carried out for the Department for Business, Energy and Industrial Strategy (BEIS) in 2021 [19]. This modelled a range of scenarios with results showing a DH potential of between 10%

and 40% of the market and with an economic potential of 20% of the heat market as a central estimate.

A report by Element Energy [20] into decarbonisation pathways for housing concluded that 5.5 m homes could be connected to DH by 2050 out of a total of 29 m, i.e., an 18% market share. In the Heat and Buildings strategy of October 2021 [2] this figure was referenced as forming the basis for a 'no regrets' policy for heat networks; page 38 of the strategy states:

> *We also believe that developing the market for low-carbon heat networks will be a no-regrets action, in recognition of the Climate Change Committee's recommendation for around 18% of UK heat to come from heat networks by 2050 as part of a least cost pathway to meeting net-zero.*

1.4 Development of district cooling

1.4.1 International development

District cooling projects were initially developed in campus applications in the USA, for example at John F Kennedy Airport with 41 MW capacity in 1967 and The World Exhibition area with 105 MW capacity in 1970.

In the USA there are now 40 DC systems supplying central business districts as well as over 150 DC systems at universities. One of the largest DC systems is in Chicago which serves 100 buildings with a capacity of 350 MW. Central chillers use ammonia refrigerant and ice storage is incorporated. Minneapolis-St Paul has an established DC network, alongside its DH network, delivering 128 GWh in 2021 [7].

In the north-west of Paris, the La Defense district was supplied with DC from 1967 initially with 87 MW later expanded to 213 MW by 2008. In central Paris, Les Halles was developed from 1978 and is now part of the Climespace system which supplied 380 GWh of cooling in 2010.

The Stockholm DC scheme is extensive and makes use of the cold water in the local fjord and 'free cooling' from large-scale heat pumps used for the DH network. It supplied 417 GWh in 2009 with total DC in Sweden now supplying 1000 GWh.

In Japan, there are many large-scale DC systems some incorporating coolth storage with a total annual delivery of 3900 GWh.

In South Korea, DC is well established with 880 MW installed capacity. Absorption chillers are also used, taking heat from the DH network to produce cooling.

The largest growth of DC has occurred in the Middle East over the last 30 years, associated with the rapid expansion of cities in UAE, Qatar, Saudi Arabia, Bahrain and Kuwait. Installed capacity in this region exceeds 9000 MW with the potential to grow to 30 000 MW. The share of the market is estimated at around 2% in the region and 10% in UAE.

For further information on the development of international DC schemes see [6, 9].

1.4.2 Development of district cooling in the UK

District cooling projects in the UK are not common partly due to the UK's relatively cool summer climate but significant examples may be found.

The Southampton scheme (1986) provides cooling to the shopping centre and other commercial buildings. Total chiller capacity is 13 MW including an absorption chiller.

The Citigen scheme in the City of London (1993) supplies DC to a number of public and commercial buildings and for a time also provided a heat rejection circuit for chillers in Smithfield Market. Initially absorption chillers were used but these were replaced later with vapour-compression machines. Total chiller capacity is now 6.3 MW.

Terminal 5 Heathrow (from 2005) has an Energy Centre that supplies cooling to the main terminal building, satellite terminals as well as ancillary buildings. The chillers have a total capacity of 35 MW (five at 7 MW) and use ammonia refrigerant and evaporative cooling towers.

At the Queen Elizabeth Olympic Park (from 2010) there is an extensive DC system supplying the sports venues and the major shopping mall as well as hotels and offices. The chilled water is produced in two Energy Centres using both ammonia chillers and double effect absorption chillers. Total chiller capacity is 53 MW (see Box 6.5 in section 6.7).

Bibliography

[1] The London Plan 2021 GLA 2021 (https://london.gov.uk/programmes-strategies/planning/london-plan)

[2] Department for Business, Energy and Industrial Strategy 2021 *Heat and Buildings Strategy* CP388 (https://gov.uk/government/publications/heat-and-buildings-strategy) 192 p

[3] Ramboll Group *District Energy* (https://ramboll.com/energy/district-energy)

[4] Euroheat and Power 2023 *District Energy Explained* (https://euroheat.org/knowledge-hub/district-energy-explained.html)

[5] DESNZ/BEIS 2022 *Energy Flow Chart for 2021* (https://gov.uk/government/statistics/energy-flow-chart-2021)

[6] Frederiksen S and Werner S 2013 *District Heating and Cooling* ch 12 Studentlitteratur

[7] District Energy St Paul *Annual Report 2021* (https://districtenergy.com/wp-content/uploads/2022/02/2021-Annual-Report.pdf 2021-Annual-Report.pdf)

[8] Ryding H 1986 The performance of Moscow's district heating system, technical and economic impact of co-generation *IMechE Seminar* (MEP)

[9] Wiltshire R (ed) 2016 *Advanced District Heating and Cooling (DHC) Systems* (Woodhead Publishing) ch 1 and 2

[10] Euroheat and Power 2022 *DHC Market Outlook* (https://euroheat.org/Euroheat&Power)

[11] Homes and Communities Agency 2011 *District Heating Good Practice: Learning from the Low Carbon Infrastructure Fund* (https://theade.co.uk/resources/guidance/district-heating-good-practice-guide-for-local-authorities)

[12] Islington Council Bunhill Phase 2 video (https://youtube.com/watch?v=unuRwEBG24I

[13] Pöyry Energy Consulting & Faber Maunsell (AECOM) 2009 The potential and cost of district heating networks *Technical Report*, Department of Energy and Climate Change, London

[14] Department of Energy and Climate Change 2013 *The Future of Heating: Meeting the Challenge* (https://gov.uk/government/publications/the-future-of-heating-meeting-the-challenge)

[15] Heat Networks Delivery Unit *Support and Guidance for Local Authorities Developing Heat Networks* (https://gov.uk/guidance/heat-networks-delivery-unit)

[16] *Heat Network Support Unit Scotland* (https://heatnetworksupport.scot/)

[17] Energy Technologies Institute 2012 *Macro Distributed Energy Summary Report*, ETI (https://eti.co.uk/library/macro-distributed-energy-project/)

[18] Ricardo AEA 2015 *National Comprehensive Assessment of the Potential for Combined Heat and Power and District Heating and Cooling in the UK* (https://gov.uk/government/publications/the-national-comprehensive-assessment-of-the-potential-for-combined-heat-and-power-and-district-heating-and-cooling-in-the-uk)

[19] Department for Business, Energy and Industrial Strategy 2021 *Opportunity Areas for District Heating Networks in the UK* (https://gov.uk/government/publications/opportunity-areas-for-district-heating-networks-in-the-uk-second-national-comprehensive-assessment)

[20] Element Energy April 2021 *Development of Trajectories for Residential Heat Decarbonisation to Inform the Sixth Carbon Budget. A Study for the Committee on Climate Change* Element Energy (https://theccc.org.uk/publication/development-of-trajectories-for-residential-heat-decarbonisation-to-inform-the-sixth-carbon-budget-element-energy/)

Paul Woods, MA MSc CEng FIMechE FEI FCIBSE

Chapter 2

Key features of DH/C systems

2.1 Component parts of a DH system

2.1.1 Introduction

A typical DH system comprises five elements as shown in figure 2.1:

- The building heating systems
- The interface equipment connecting the building heating system to the heat network
- The heat network of flow and return insulated pipes
- The primary heat source which is typically located in an 'energy centre'
- Ancillary plant including peak and standby boilers, thermal energy storage and DH pumps, normally within or adjacent to the energy centre but can be within the network

Each of these elements are discussed in turn below.

2.1.2 Building heating systems

It may seem inappropriate to include the building heating system as part of the DH system, however, it is important that the design and operation of the building's heating system is compatible with and optimised to work with the DH system. This can be a challenge when connecting existing buildings where it may be impractical or costly to change the existing heating systems and a compromise position may need to be found. However, for new buildings, the building's heating system can be designed to be optimal for the DH system. Generally, any existing heating system with a centralised source for the building and using water as a heat transport and heat transfer medium (i.e., a 'wet heating system') can be connected to DH whether using radiators, fan-coil units, air handling units or underfloor heating. The main compatibility issues are the operating temperatures of the building's heating system and the type of controls used for the space heating and domestic hot water circuits.

doi:10.1088/978-0-7503-5286-4ch2

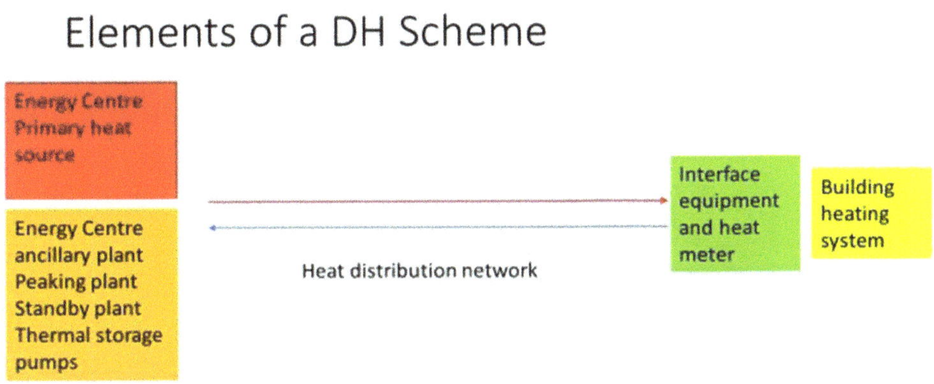

Figure 2.1. Elements of a DH system.

Chapter 3 discusses these issues in more detail.

2.1.3 Interface equipment

The interface equipment needed at the point of connection to the building heating system is a relatively small part of the total system cost but can have a major impact on the efficiency of the network. As a minimum, there will be facilities for: isolation of the building from the network for maintenance purposes, a method of limiting the flow rate from the network to an agreed maximum level and a heat meter to measure the amount of heat supplied. In addition, there may be controls to set the temperature of the supply to the building and one or more heat exchangers to physically separate the water in the DH circuit from the water in the building's space heating and hot water heating system. The contractual interface with the customer for the heat supply is normally at the point of connection to the building but the ownership and maintenance of the interface equipment itself varies. In some countries the connection equipment is the responsibility of the customer whereas in the UK it is more usual for the DH company to own the equipment.

For dwellings, the interface equipment is normally pre-assembled and supplied as a heat interface unit (HIU) which is of similar size and appearance to a domestic gas boiler (see figure 2.2).

For non-residential buildings the interface equipment (often termed a thermal substation) is also often pre-assembled and factory tested (see figure 2.3). This unit is much more compact than the boiler plant it would be displacing and, if these boilers are removed after connection to DH, the customer will gain additional space within their building.

Heat meters are available which measure the flow rate and the flow and return temperatures at the supply point and from this the heat energy taken by the customer is automatically calculated and displayed on the meter. Automatic meter reading systems are available so there is no need to visit the property to take the meter reading. The data on flow rate and temperatures is also useful for monitoring the performance of the scheme (see section 8.3).

Chapter 4 describes the interface equipment in more detail.

Figure 2.2. Typical Heat Interface Unit for dwellings (without cover and with cover) (photos courtesy Switch2 Energy Ltd). Reproduced with permission from Switch2 Energy Ltd.

Figure 2.3. Typical thermal substation (photo courtesy Heatweb Solutions). Reproduced with permission from Heatweb Solutions.

2.1.4 Heat network

The heat network consists of a pair of insulated pipes, one taking the higher temperature water from the energy centre (the flow pipe) and the other returning cooler water back to the energy centre after the heat has been delivered into the building (the return pipe). It is a closed system with no water loss. The pipes are sized to deliver the required flow rate in the most economic way, by balancing the initial cost with the cost of pumping energy and heat loss.

The heat energy transported in the pipe is a function of the flow rate and the temperature difference between flow and return. So, for a given heat demand, a larger temperature difference enables flow rates to be reduced and smaller pipe sizes selected.

Since the 1970s pre-insulated pipes have been used which can be buried directly in the ground as for other piped utilities such as gas and water (see figures 2.4 and 2.5). The construction comprises:

Figure 2.4. Pre-insulated steel pipes. Reproduced with permission from Vital Energi.

Figure 2.5. Installing pre-insulated steel pipes in a trench. Reproduced with permission from Vital Energi.

- A carrier pipe in which the water flows, which can be steel or plastic
- An insulation layer which may be rigid polyurethane foam or a more flexible foam material
- An outer casing to protect the insulation and carrier pipe, typically high density polyethylene

The carrier pipe is joined by welding or, in the case of some types of plastic pipe, mechanical couplings. The joint area is then closed up using a pipe sleeve which is welded to the outer casing and insulation is injected into the void left at the joint.

Systems with steel pipes are very robust and can have a long life. Pipes are supplied in straight lengths of 6, 12 or 16 m but they can also be supplied as factory made curved pipes. The main cause of failure is corrosion from either the inside or outside of the pipe. Corrosion from the inside is prevented by close control of the water quality used in the DH circuit. External corrosion depends on maintaining the integrity of the outer casing and preventing ground water penetration, especially at the joints. Alarm wires embedded in the insulation are used to detect the presence of moisture within the insulation layer so that a repair can be carried out before any significant pipe corrosion has occurred. Steel pipes constrained in the ground and then heated will have significant compressive stresses so expansion loops or bends may be needed to limit stress levels.

Plastic carrier pipes are lighter and have the benefit of flexibility. This means they can be more easily installed around obstacles in the ground or where ground levels vary. They can also be supplied as coils in long lengths of more than 100 m, which means fewer joints are required (see figure 2.6). Expansion bends or loops are not needed as the expansion stresses can be absorbed within the material. Most plastic pipe systems have been supplied for the smaller diameters especially where long lengths are needed without branches. However, if required, larger pipes are supplied as straight lengths similar to the steel pipes.

Although two separate pipes for flow and return are commonly used, a more efficient solution in terms of limiting heat loss is to have two carrier pipes in one casing (see figure 2.7). These 'twin pipes' are particularly used for smaller diameters and where changes in the installation level are not significant and are available for both steel and plastic carrier pipes. Welding of the steel twin pipes is a skilled task due to the restricted access (see figure 2.8).

Figure 2.6. Pre-insulated pipes with plastic carrier pipe. Reproduced with permission from REHAU Ltd.

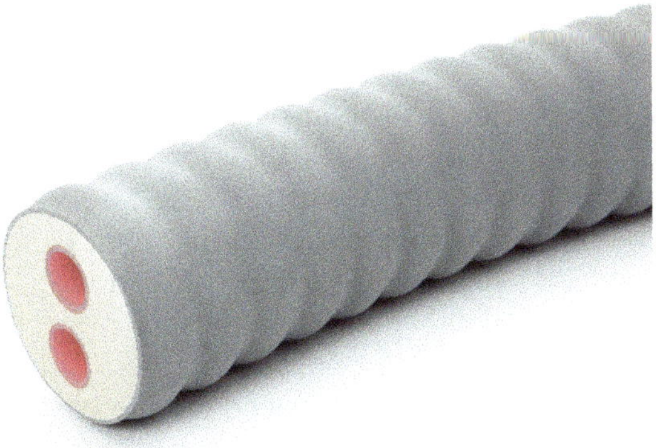

Figure 2.7. Pre-insulated plastic twin pipes. Reproduced with permission from REHAU Ltd.

Figure 2.8. Installing pre-insulated steel twin pipes. Reproduced with permission from Vital Energi.

Chapter 5 provides more detailed information on the design of heat networks and the technologies used.

2.1.5 Primary heat source and energy centre

It is the primary heat source that delivers both the economic case for DH and its environmental benefits. There is usually only a small energy efficiency benefit from DH if the heat source is the same as that which can be used in individual buildings (although there may be a cost benefit due to economies of scale). The primary heat source needs to provide a low cost and low carbon solution. A wide range of heat sources can be used and these have been grouped under four broad headings below:

(A) Fuels that would be environmentally damaging, difficult or impossible to use in individual buildings, e.g., domestic and commercial waste, biomass, heavy fuel oil (HFO) and coal. UK examples are:

- Waste boilers: Sheffield (first phase), Lerwick
- Waste CHP: London SELCHP (see figure 2.9), Sheffield, Nottingham, Coventry, Cardiff, Leeds
- Biomass boilers: Queen Elizabeth Olympic Park, Bristol
- Biomass CHP: Sheffield Blackburn Meadows
- HFO boilers: 1960s and 1970s housing projects e.g. in Southwark, Tower Hamlets
- Coal boilers: Doncaster, Bridgend (in the past)

(B) Heat derived from processes which are more efficient and/or more economic at a district scale than individual buildings, e.g., combined heat and power (CHP), or where the source is only available in bulk at a

Figure 2.9. The SELCHP Energy-from-waste CHP plant, London File:SELCHP.jpg - Wikimedia Commons. This SELCHP.jpg image has been obtained by the author(s) from the Wikimedia website where it was made available under a CC BY-SA 3.0 licence. It is included within this book on that basis. It is attributed to Pixel8.

particular site, e.g., industrial heat recovery. There are many UK examples using CHP including:

- Spark-ignition gas-engines (see figure 2.10): Pimlico District Heating Utility, Queen Elizabeth Olympic Park, Kings Cross, Bunhill Phase 1 (Islington), Bristol, Gateshead, Edinburgh University
- Dual-fuel engines: Southampton, Citigen (City of London)
- Gas turbines: Whitehall

(C) Renewable heat sources which are only economic at a district scale, e.g., deep geothermal, solar thermal with interseasonal storage
Examples:

- Deep geothermal: Southampton
- Solar thermal: Marstal, Vojens, Denmark (see figure 2.11) Greening the heating supply through the use of thermal solar storage (stateofgreen.com)

(D) Heat pump systems where there is a heat source with a suitable elevated temperature available, e.g., a major datacentre, an industrial process with heat rejection, a river or minewater. Examples are:

- Tallaght, Dublin (using heat from a datacentre)
- Islington (using warm air from underground railway ventilation)
- Castle Park Bristol (using harbour water) see figure 2.12
- Seaham (using minewater—in development)

Figure 2.10. Spark-ignition gas-engine CHP. Reproduced with permission from Vital Energi.

Further information on the design of primary heat sources is given in chapter 6.

2.1.6 Ancillary plant

In addition to the primary heat source the Energy Centre normally contains four other main items of ancillary plant: peaking plant, standby plant, thermal storage and DH pumps.

Peaking plant

The DH system needs to be designed to meet the heat demand on the coldest days of the year. But these periods are of relatively short duration and occur rarely over the year. It is usually more economic to size the primary heat source at a smaller capacity than this peak demand and use lower capital cost heating plant to operate in the peak periods to supplement the main plant. Even if this peaking plant has a higher CO_2 emission and higher operating costs this will not have a large impact if it is only used to meet a small proportion of the annual heat supply.

Figure 2.11. District heating scale solar thermal plant at Marstal, Denmark File:Marstal.powerplant.1.jpg - Wikimedia Commons. This Marstal.powerplant.1.jpg image has been obtained by the author(s) from the Wikimedia website where it was made available under a CC BY-SA 3.0 licence. It is included within this book on that basis. It is attributed to Erik Christensen.

Figure 2.12. Water source heat pump at Castle Park, Bristol.

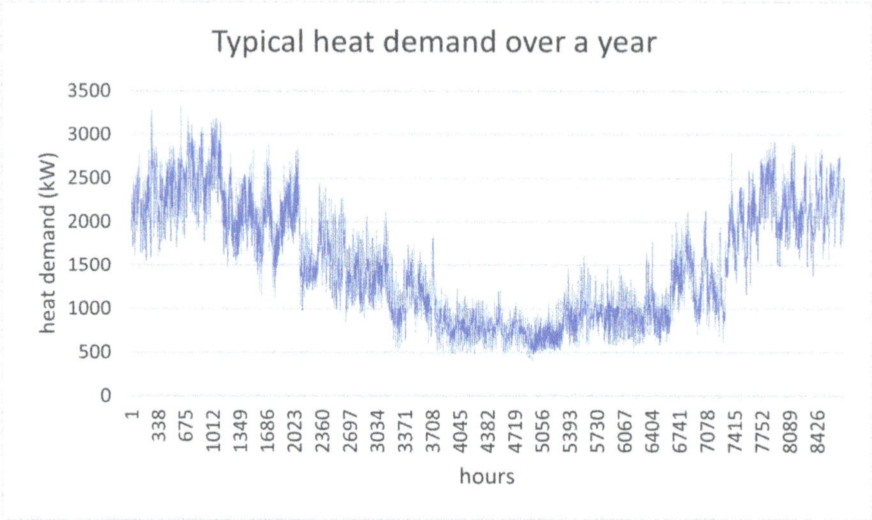

Figure 2.13. Typical variation of heat demand over a year.

This can be illustrated using a typical hourly demand pattern for a whole year (see figure 2.13).

A useful parameter is the load factor which is defined as:

$$LF = \text{Annual heat demand (MWh)/peak demand (MW)} \times 8760$$

The load factor for this set of data is 46% which is typical for a scheme dominated by housing and including heat losses.

If the same data is plotted by sorting the data from highest to lowest demand a 'load duration curve' is created which can be used to show how multiple heat sources can meet a varying heat demand (see figure 2.14). This diagram is important for the designer as it shows that for most of the time the DH system will be operating to deliver demands much lower than the peak demand. There is very little benefit in sizing the low carbon plant at more than about 50% of the peak demand. Also, the load duration curve shows that designing for efficient operation at part-load is important as for over half the year the demand is less than 50% of the peak.

A plot of the amount of the annual demand that can be supplied by varying the supply capacity is also useful. This is effectively the area under the load duration curve (see figure 2.15). This shows that a primary heat source sized at 50% of the peak demand would supply about 85% of the total annual heat demand and a primary heat source of 40% of the peak demand would supply about 75% of the total annual heat demand. The remaining heat demand would be supplied from the secondary peaking plant.

Standby plant

The reliability of the heat network needs to be very high given the number of customers impacted if there is a failure to meet demand. Most energy production

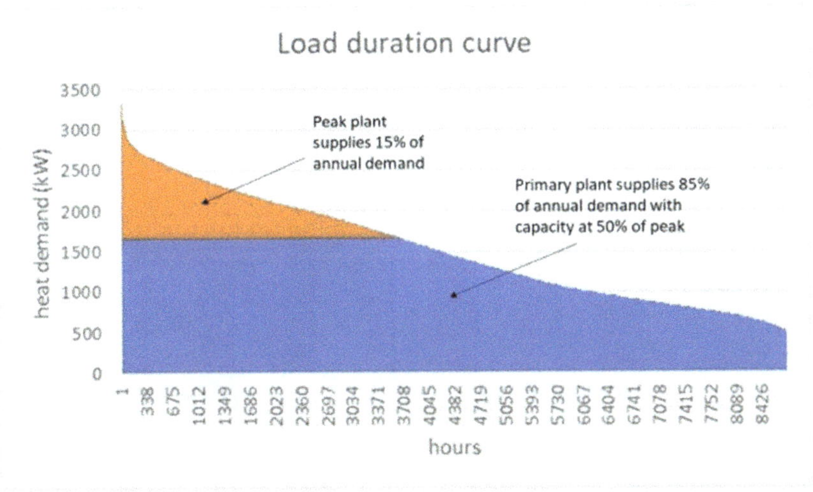

Figure 2.14. Load duration curve showing contribution from primary source and peaking plant.

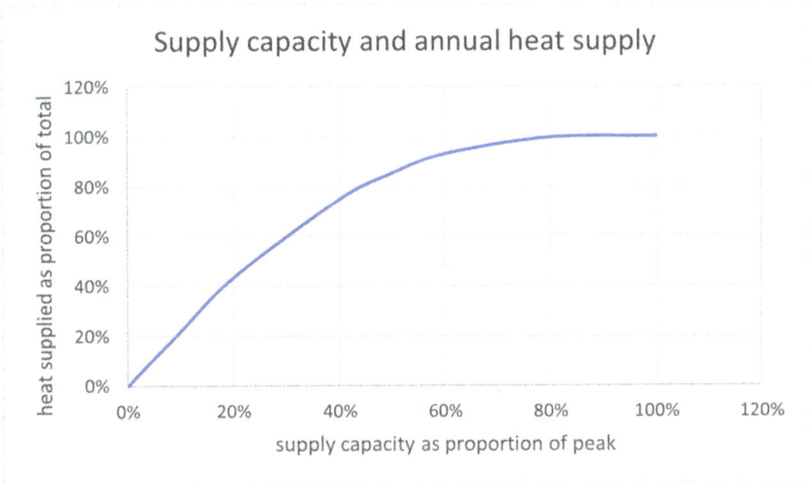

Figure 2.15. Amount of heat that can be supplied with varying supply capacity.

units will need to be shut down for planned maintenance and may also suffer from an unplanned failure. To continue to meet the heat demand in these situations, a separate standby plant is used. The capacity of this standby plant will depend on the number and type of the primary plant, the likelihood of failure, the shape of the load duration curve and the desired reliability of the heat supply. If there is a single unit for the main plant then often the standby plant is sized to meet the peak demand assuming the main plant is not operating. This capacity is more than that needed for the peaking requirement discussed above and the secondary plant may then be referred to as peaking and standby plant as it will carry out both functions.

Figure 2.16. The peak boiler plant at Saxton Gardens, Leeds. Reproduced with permission from Vital Energi.

For example, a scheme with a peak demand of 10 MW may be supplied from an energy centre containing a 5 MWth CHP plant and 10 MW of gas-fired boilers. The gas-fired boilers are operated both to meet peak demands and to provide standby on the CHP plant. In other cases, however, the peak boilers and standby boilers may be of different types and capacities.

Peak and standby boilers are usually located with the primary heat source in an energy centre, however, they can also be installed elsewhere on the network, especially where the primary heat source is some distance from the heat demand. The transmission pipe from the primary heat source to the network can then be sized just to deliver the primary heat source capacity and not the peak demand thus reducing its cost. This was the solution adopted for the Leeds PIPES project (see figure 2.16).

Thermal energy storage

One of the benefits of a DH system is the ability to incorporate thermal energy storage (TES) at a scale that would be impractical at a building level especially within existing buildings. Most thermal stores consist of a cylindrical steel storage vessel orientated vertically. This remains full of water but the temperature of the water in the store varies between the network flow temperature and the return water temperature. With sufficient temperature difference the hot flow water floats above the colder return with limited mixing. The amount of energy in the store depends on the proportion of hotter to colder water.

TES decouples the heat demand from the heat production and so offers a number of advantages including:

Figure 2.17. Photo of thermal store in operation in the Coventry district heating scheme.

- Reducing the peak demand imposed on the production plant as short-term peaks can be supplied from the TES, leading to a lower installed capacity and less use of peaking boilers
- Enabling the main heat source to operate at times which minimise the cost of operation, e.g., a CHP plant can be operated mainly at times of high electricity price, and a heat pump operated mainly at times of low electricity price rather than having to operate to follow the heat demand pattern.
- Enabling the main heat source to operate at levels of output that maximise efficiency rather than operating to follow the heat demand pattern, e.g., the CHP plant could be operated at full output for two periods in the day rather than continuously at varying loads.

With the increased contribution from variable renewable electricity sources such as wind and solar, the variation in electricity prices is likely to become more pronounced and with the variation not just reflecting a diurnal pattern but with longer periods of higher or lower prices depending on wind strength. This will mean that TES is likely to become more important in the future as a means of managing renewable electricity generation on the national electricity grid and obtaining additional financial benefits (see section 10.2).

Although TES is most commonly located at the energy centre it can also be installed within the heat network at strategic points, e.g., in Coventry the TES is within the city not at the energy centre which is a waste to energy plant some 2 km away, see figure 2.17.

DH pumps

Pumps need to be installed within the network to provide the pressure required to create the flow in the network. These DH pumps are normally located at the energy centre where they can be easily monitored and maintained although sometimes additional pumps are required within the network. Standby pumps are provided so that should a pump fail a standby one can be automatically started to maintain the supply of heat. Multiple pumps may be used for more efficient operation so that a smaller pump can be used at times of part-load. Pumps normally have variable speed drives to improve efficiency further so that the pressure created matches more closely the requirements of the network.

Chapter 6 provides more information on heating production plant.

2.2 Component parts of a DC system

A typical district cooling system is similar to the DH system comprising:

- The building cooling systems
- The connection between the building cooling system and the district cooling network
- The district cooling network of flow and return insulated pipes
- The coolth source which may be located in an energy centre
- Ancillary plant, including TES and DC pumps

There are important differences between DH and DC however and DC faces a number of challenges when compared to DH.

In most northern climates the need for cooling is restricted to relatively short periods in the year except for spaces with limited ventilation and high internal gains such as highly serviced office buildings with high occupancy. So, any DC system needs to recover its costs in relatively few operating hours in the year which will result in a higher cooling energy price. DC is, therefore, more likely to be used in higher density areas and over a limited distance.

A cooling network (using water) has a restricted range of temperatures in which to operate. It needs to be above 0 °C to avoid freezing and below the required room temperature of 20 °C. This means that flow rates and pipes are larger and more costly. Pipes will still be insulated but not to the same extent as for DH pipes.

As for a heating system, the building's cooling systems need to be designed to work with appropriate temperatures and controls. For DC a high return temperature is desirable which might be obtained where, e.g, chilled beams are used which need to be operated with higher chilled water temperatures.

The building connection equipment needs to be carefully designed to minimise temperature differences between the network and the building's cooling system.

Although heat gains are low from the heat transfer with the ground, the impact of pumping energy is more important as this energy leads to unwanted heating of the circulating water.

It may be difficult to achieve a significant cost difference in capital or operating costs between centralised chillers and local chillers. However, there are some potential advantages with centralised chillers in addition to economies of scale:

- Thermal energy storage can be incorporated allowing electricity to be purchased at lower prices at night for example (see section 6.14).
- Alternative refrigerants can be used, e.g., ammonia which may not be acceptable within buildings due to safety concerns; ammonia has zero global warming impact.
- Evaporative cooling towers can be used to improve efficiency which may be more difficult to integrate into an existing building due to the visual impact of vapour plumes and safety concerns regarding legionella.
- Linking DH and DC together is possible in various ways to improve efficiency (see section 10.7).

Section 6.13 provides further information on DC production plant.

2.3 Trends in DH design

With the perspective of over 50 years the industry has described the development of DH design in terms of 'generations' as illustrated in figure 2.18 [1].

First generation systems used steam as the heat transport medium as in Manhattan, New York and Copenhagen.

Second generation systems generally used large-scale power stations as heat sources with flow temperatures around 120 °C at peak. Many of these systems are

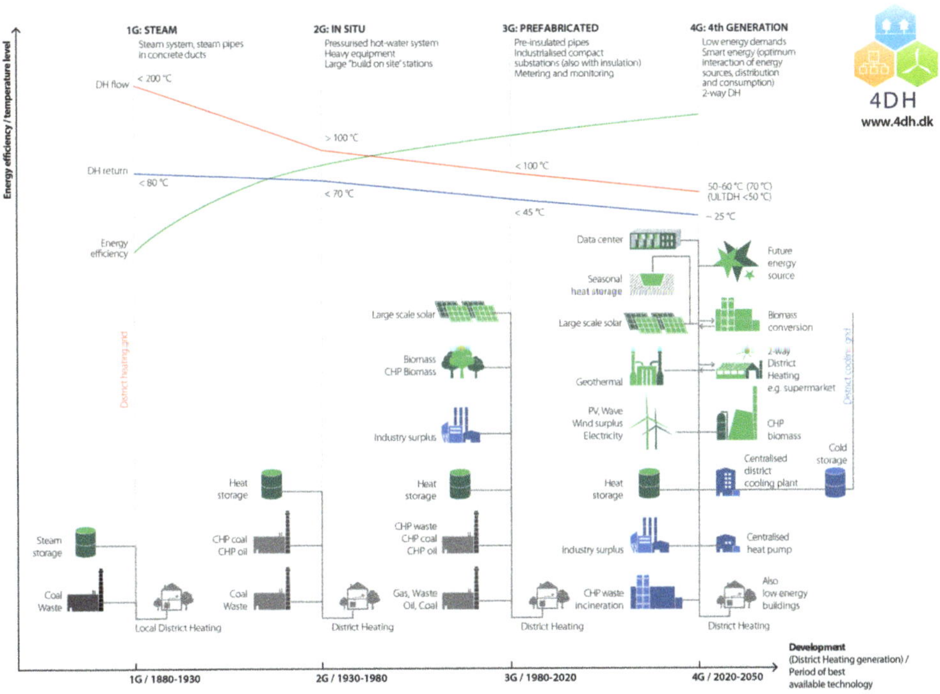

Figure 2.18. The development of DH from 1G to 4G systems. Reproduced with permission from [1].

still in operation but the flow temperature is usually reduced as far as possible to suit the demand.

Third generation systems were often smaller and used gas-engine CHP operating with flow temperatures below 95 °C.

Fourth generation systems use lower temperatures, partly to reduce heat losses from the network and partly to enable greater use of heat pumps, heat recovery and renewable sources. Heat pump efficiency will be higher if supplying at a lower temperature. Fourth generation systems include significant TES and operate in a flexible way to provide additional system benefits to the electricity grid. The definition of fourth generation is, therefore, not only about operating temperatures.

2.4 Ambient temperature networks

Sometimes referred to as fifth generation heat networks, or cold networks, ambient temperature networks distribute the water near to ambient temperature. The provision of useful heat for the building heating system is achieved by heat pumps in each building. The ambient network can also be used as a method for heat rejection from buildings using cooling plant. This enables heat rejected from one building that requires cooling to be used as a heat source for another building that requires heating.

The main advantage is that heat losses from the network are much less as the operating temperature is close to ambient. However, having heat pumps within each building involves a loss of useful space, especially in dwellings as a hot water cylinder will be required, and may be more expensive than using centralised heat pumps.

Ambient networks are discussed further in section 10.8.

Bibliography

[1] Thorsen J E, Lund H and Mathiesen B V 2018 *Progression of District Heating—1st to 4th Generation* (https://vbn.aau.dk/en/publications/progression-of-district-heating-1st-to-4th-generation-3)

Part II

The Engineering of DH and DC systems

IOP Publishing

An Introduction to District Heating and Cooling
Low carbon energy for buildings
Paul Woods, MA MSc CEng FIMechE FEI FCIBSE

Chapter 3

Building services—heating and cooling systems

3.1 Importance of a building's heating and cooling systems

The efficient operation of a DH/C system depends significantly on the design and operation of the heating and cooling systems within the connected buildings (the building services). When connecting existing buildings, the DH/C company may have little influence on how the building is operated except through the heating or cooling energy sales contract. For new buildings, the DH/C company may be in a better place to influence the design by engaging with the building's designers so that they can take account of the technical requirements of the DH/C system.

The three energy requirements are:

- Space heating
- Domestic hot water heating (DHW)
- Space cooling

For each of these, the main areas of concern relate to:

- Assessment of peak demands
- Equipment needed for the heat transfer to the space or domestic hot water circuits
- Operating temperatures
- Methods used for control of space and DHW temperatures

In all of these aspects, the DH/C designer will need to consider the impact of the building services design and operation on the cost and efficiency of operation of the DH/C system. There is a need to consider: network capital cost, heat loss/gain, pumping energy and in relation to the DH/C heating and cooling plant, its capital cost and operational efficiency.

doi:10.1088/978-0-7503-5286-4ch3

Designers often focus on the peak load condition, which needs to be met to ensure comfort on the coldest or hottest days, but these occur rarely. When designing for the lowest lifecycle cost the efficiency at more typical operating conditions, when the heating or cooling demands are far lower than the peak, needs to be considered. Part load and no load are the predominant operating conditions for space heating or cooling and DHW systems and efficient operation of the DH/C system needs to be achieved under these conditions.

This chapter discusses each of these aspects with reference to the engineering principles involved. The final section contains recommendations for making modifications to building services in existing buildings to be connected to DH/C that will improve network performance.

3.2 Space heating

3.2.1 Assessing peak demands

The peak demand of a building to be connected to DH needs to be established without overestimating. The peak demand will be used to size the building connection equipment, the heat network and the total capacity of top-up and standby plant. It is necessary to agree with the customer the external temperature conditions that relate to the design peak. For existing buildings, whenever possible, it is worthwhile to establish the actual peak heat demands from heat meter data or from half-hourly gas meter data. For conventional heating using gas-fired boilers an over-estimate of the peak does not result in a large capital cost penalty but with DH the capital cost is much more significant and the majority of this cost is related directly to peak demand.

3.2.2 Heat transfer equipment

For space heating, steel panel radiators are commonly used in residential buildings and for perimeter heating in non-residential buildings and these can be supplied from DH systems provided they are balanced correctly. These transfer heat to the space partly by radiation but mainly by the natural convection of air across the surface and attached fins. As a result, the output of the radiator is very dependent on the operating temperatures (see section 3.2.3). The pressure rating of a radiator can be lower than that of the connecting pipework and this is important to consider if the direct connection method is to be used (see section 4.2).

Fan-assisted radiators are becoming more commonly used in systems using heat pumps as they can operate with lower temperatures and these can also be used to advantage with DH.

Underfloor heating has been less commonly used in the UK than elsewhere but is becoming more popular for new-build homes. It is also compatible with DH and offers some advantages as operating temperatures are lower than for radiators.

Air handling units are mainly used in non-residential buildings and because the air is heated directly operating temperatures can again be lower than with radiators.

Fan-coil units are commonly used in non-residential applications often as combined heating and cooling units. These can operate with low temperatures as they rely on forced convective heat transfer to the air. They are increasingly used for new-build residential developments especially where cooling is also provided.

Heating systems that are not compatible with DH and would need replacement include: direct electric heating, electric storage heaters, gas-fired warm air units or AHUs, and gas-fired radiant heating. Also, high temperature radiant panels using steam or high temperature hot water would not be suitable.

3.2.3 Operating temperatures

A designer will need to size the heat emitters to meet the heat demand of the space for a given external design air temperature. To do this the operating temperatures of the heating circuit need to be defined.

Traditionally in the UK flow and return temperatures of 82 °C/71 °C (180 °F/160 °F) were used. These were the temperatures that were needed to avoid corrosion in domestic coal-fired back boilers which occurs due to acidic water condensing from the flue gases onto the heat exchanger surfaces. With the use of natural gas as a cleaner fuel and the development of condensing boilers these temperatures are no longer relevant and lower temperatures are commonly used to select radiators, especially for new-build properties or where heat pumps are used. New-build properties will have relatively small heat losses and so the larger size of radiator needed for the lower temperatures will not be a major issue in terms of space requirement. The Building Regulations (England) now require heating systems in new buildings and completely new heating systems in existing buildings to be designed with a maximum flow temperature of 55 °C; this is so that they will be compatible with heat pumps in the future [1]. This is likely to mean that underfloor heating will become more prevalent in future new-build properties to avoid the disadvantages of using large radiators or fan-assisted radiators.

Traditions and standards in other countries differ—in Sweden for example temperatures of 60 °C/40 °C have been required for radiator systems to be connected to DH since 1982, partly to promote the greater use of low temperature heat sources [2].

Since 2000, the standard BS EN 442 has been used by most manufacturers when stating radiator outputs. These are based on a 50 °C temperature difference between the mean radiator temperature and the room temperature, e.g., 80 °C flow 60 °C return with a room temperature of 20 °C.

When connecting existing buildings to DH it will be important to minimise cost and disruption and so retaining the existing heat emitters would be preferred. Consideration needs to be given as to how the peak heat demand is to be met and whether the operating temperatures can be modified. For example, it may be beneficial for the heat network to operate with lower return temperatures to minimise pipe sizes and heat losses or to operate with lower flow temperatures where heat pumps are used as the main heat source.

The amount that operating temperatures can be changed depends on the degree of oversizing that may be present in the existing heating systems. Heat emitters are typically oversized because:

- Internal gains are not normally taken into account when sizing
- Design margins added may be excessive
- Building fabric improvements may have been made since the original selection was made, e.g., additional insulation, new windows, draught-stripping of doors

The impact of changes to operating temperatures on heat emitter outputs can be established by consulting manufacturers' catalogues where outputs for a range of temperatures may be given. The outputs also depend on the connections to the radiator. The preferred approach to minimise return temperatures for a given size and required output is top entry, opposite bottom exit.

A commonly used equation [3] for the output of a radiator is:

$$Q = k\ A\ (T_m - T_a)^{1.3} \tag{3.1}$$

where:

Q is the output of the radiator (W)
k is a constant
A is the area of the radiator
T_m is the mean temperature of radiator, i.e., $(T_f + T_r)/2$, where:
T_f = flow temperature, T_r = return temperature
T_a = air temperature in the heated space

Using this equation, the output of a radiator expressed as a ratio to its output at 80 °C flow 60 °C return is given in table 3.1.

Figure 3.1 below shows how the radiator output drops with reductions in return temperature for a constant flow temperature of 80 °C and figure 3.2 shows the relation between flow and return temperatures for a constant output (in this case set at 75% of the 80 °C/60 °C output).

Table 3.1. Output of radiators for a range of flow and return temperatures relative to 80 °C/60 °C standard.

Flow temperature (°C)	50	55	60	65	70	75	80	85
Return temperature (°C)								
30	0.30	0.35	0.41	0.46	0.51	0.57	0.63	0.69
35	0.35	0.41	0.46	0.51	0.57	0.63	0.69	0.75
40	0.41	0.46	0.51	0.57	0.63	0.69	0.75	0.81
45	0.46	0.51	0.57	0.63	0.69	0.75	0.81	0.87
50		0.57	0.63	0.69	0.75	0.81	0.87	0.94
55			0.69	0.75	0.81	0.87	0.94	1.00
60				0.81	0.87	0.94	1.00	1.07
65					0.94	1.00	1.07	1.13
70						1.07	1.13	1.20

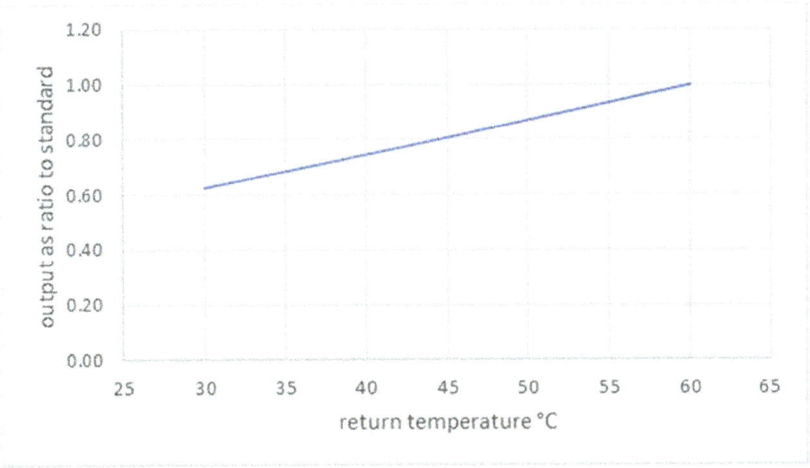

Figure 3.1. Output of radiator with varying return temperature and a constant 80 °C flow temperature.

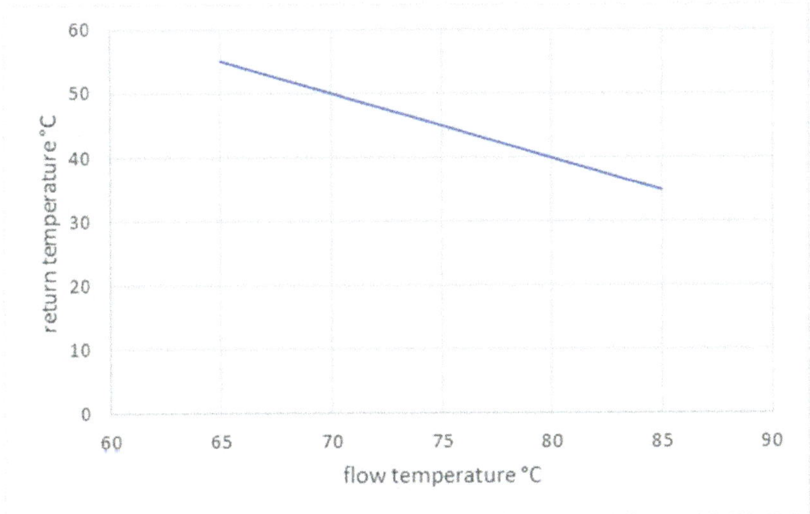

Figure 3.2. Relation between flow and return temperatures for constant heat output.

If the original selection of temperatures is assumed to be 80 °C/60 °C, then the operating temperatures in table 3.2 will still deliver 87% of the original output which is likely to be sufficient to meet the demand on the peak design day in most cases. A survey of 500 dwellings [4] showed that 63% had heating systems able to meet the peak demand with a mean temperature of 65 °C or less (e.g. 80 °C/50 °C).

Table 3.2. Flow and return temperatures to deliver 87% of standard output.

Flow temperature (°C)	Return temperature (°C)
85	45
80	50
75	55
70	60

If a large temperature difference across the radiator is required to reduce return temperatures it is necessary to have low flow rates. Pre-settable thermostatic radiator valves (TRVs) designed specifically for low flow rate applications should be selected. For example, the Oventrop AF valve [5] with a pressure difference of 0.2 bar would require a setting 2 (out of 6) and result in a flow rate of 0.0065 l s^{-1} which corresponds to 0.82 kW radiator output for a temperature difference of 30 °C. Another example is the Danfoss RA-DV valve [6] which would use a pre-setting of 3 (out of 7) for the same flow rate and pressure difference. Pressure-independent TRVs provide a greater accuracy in flow rate setting as the flow rate is maintained over a range of pressure differences.

Flow rates less than 0.0065 l s^{-1} that may be needed for smaller radiators may be difficult to achieve with any accuracy. The opening in the valve to limit the flow to such low flow rates is very small and is, therefore, prone to blockage by debris in the system. For practical reasons a temperature difference of less than 30 °C is often preferred.

For other heating equipment such as AHU heating coils and fan-coil units where the heat transfer is by forced convection the standard heat transfer equation can be used [7] (see figure 3.3).

$$Q = UA\ LMTD \tag{3.2}$$

where:

Q is the rate of heat transfer (W)

U is the overall heat transfer coefficient; within a certain range of flow velocities this can be assumed constant

A is the area of the heat exchanger surface

$LMTD$ is the log-mean temperature difference defined as:

$LMTD = \Delta T_o - \Delta T_i / \ln(\Delta T_o / \Delta T_i)$ where

ΔT_i is the difference between primary and secondary temps at the primary entry point, $(T_1 - T_4)$

ΔT_o is the difference between primary and secondary temps at the primary exit point $(T_2 - T_3)$

In addition to equation (3.1), Q must also equal the heat lost from the primary fluid and the heat received by the secondary fluid (neglecting any heat losses from the heat

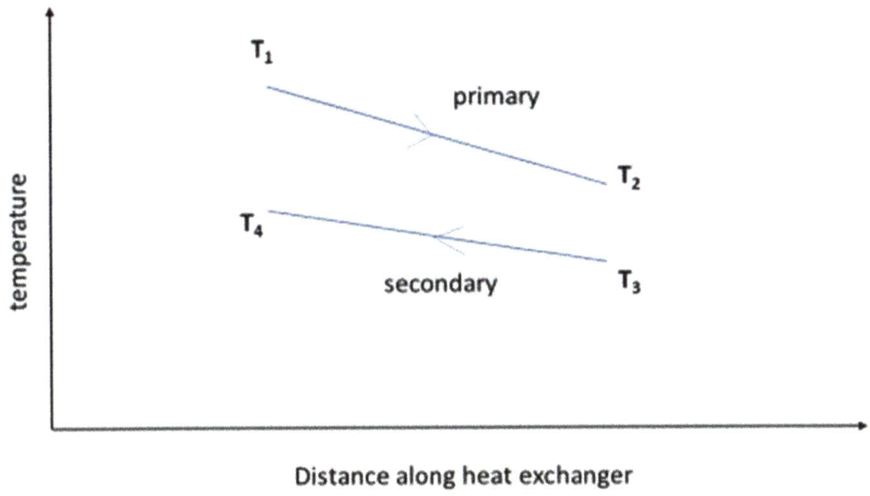

Figure 3.3. Diagram of heat exchanger temperatures (counterflow).

exchanger itself). Hence for any heat exchanger there are three equations for Q that must be satisfied:

$$Q = UA\ LMTD$$
$$Q = m_p\ c_p(T_1 - T_2)$$
$$Q = m_s\ c_s(T_3 - T_4)$$

where m_p is the mass flow rate of the primary fluid and c_p is its specific heat; where m_s is the mass flow rate of the secondary fluid and c_s is its specific heat; temperatures T_1, T_2, T_3 and T_4 are as defined on figure 3.3.

Three types of calculation are often carried out by the designer:

(1) Sizing the heat exchanger for the required heat transfer rate Q and with defined temperatures; in this case UA is the unknown.
(2) Determining the heat transfer rate Q for a change in primary flow rate (i.e., the operation of a two-port control valve) where the heat exchanger size, primary flow temperature and secondary flow rate are unchanged; in this case the unknowns are Q, T_2 and T_4 and the three equations need to be solved simultaneously.
(3) Determining the heat transfer rate for a change in primary flow temperature where the heat exchanger size, primary flow rate and secondary flow rate are unchanged; in this case the unknowns are again Q, T_2 and T_4 and the three equations need to be solved simultaneously (see box 3.1).

Box 3.1. Worked example for a reduction in flow temperature from 85 °C to 75 °C resulting in a reduction in heat transfer from 1000 to 715 kW. This problem can be solved numerically or by trial and error within an Excel sheet.

Change in primary flow temperature		
base case:		
Primary flow temp T1	85	C
Primary exit temp T2	55	C
Secondary flow temp T3	75	C
Secondary exit temp T4	50	C
T1-T4	10	C
T2-T3	5	C
LMTD	7.21	C
c	4.18	kJ kgK^{-1}
Q	1000	kW
UA	138.63	kW K^{-1}
mp	7.97	kg s^{-1}
ms	9.57	kg s^{-1}
Lower primary flow temp T1 to:	75	C

Iteration on T2 and T4 to calculate Q	T2	T4	T1-T4	T2-T3	LMTD	Q	Q	Q
						UA LMTD	mp*c* (T1-T2)	ms*c* (T4-T3)
1st iteration	55	65	10	5	7.21	1000	667	600
2nd iteration	54	67	8	4	5.77	800	700	680
3rd iteration	53	68	7	3	4.72	654	733	720
4th iteration	53.55	67.85	7.15	3.55	5.14	713	715	714

Although rebalancing circuits to use lower flow rates and hence achieve lower return temperatures will be possible with AHU coils there are practical limits due to the thermal stresses that will result.

Impact of operating temperatures on DH system

The DH system will need to be designed to meet the demand of the buildings under the design conditions and so the building space heating and hot water heating (see section 3.3.3) system temperatures will effectively determine the minimum operating temperatures for the DH network.

From figure 3.2 above it can be seen that a higher flow temperature allows a lower return temperature to be used which is usually beneficial for DH. However, a higher flow temperature may increase the cost of heat production from some sources, e.g., heat pumps (see also section 7.4).

In summary, a robust approach for existing heating systems to be compatible with DH would be to install pressure-independent TRVs set up to give radiator circuit temperatures of around 80 °C flow 50 °C return which would provide 87% of the original design output assuming this was based on 80 °C/60 °C temperatures. With more detailed information lower temperatures such as 75 °C/45 °C could be used giving 75% of the output at 80 °C/60 °C.

Operating temperatures are discussed further in chapter 7 which considers the issue of optimal selection of operating temperatures for the whole system.

3.2.4 Control of space temperatures

Section 3.2.3 considered the selection of temperatures for the design case, i.e., peak demands and this will have an important impact on the design of the heating and cooling networks, their sizing and capital costs.

In terms of operating costs and efficiency, the way in which the building heating and cooling systems are controlled under part-load conditions is very important.

Consider a hot water circuit supplying a heat emitter which may use natural convection or forced convection to transfer heat to the air. The heat transfer is determined by the equation on the water side:

$$Q = (V/\rho)\, c\, \Delta T \tag{3.3}$$

To control the room temperature the heat transfer rate Q needs to be reduced either by reducing V or ΔT or a combination of both.

There are four methods that can be used. Figures 3.4–3.8 show the principles and also the return temperature characteristic.

1. On/off control

   A common simple method—when a set temperature of the air is reached a thermostat operates a two-port valve to turn off the flow completely and opens it again when the temperature falls below the set point. During the off period the water in the heat emitter will cool down. The return temperature will be at the design condition apart from a short period when the cooled water in the heat emitter enters the heating circuit.
2. Lower volume flow to heat emitter using diverting valve

   In this method, when the set temperature is reached the flow is diverted away from the heat emitter by means of a three-port diverting valve so that less flow enters the heat emitter. The return temperature rises as some of the flow temperature mixes with the return from the heat emitter. There is constant flow rate in the heating circuit but the ΔT reduces with load with a rising return characteristic.
3. Constant volume flow with reducing flow temperatures using mixing valve

   In this method, a mixing circuit is employed so that the flow temperature onto the emitter is reduced. The return temperature will be lower and the volume flow rate in the main circuit is also reduced. Three-port control valves can be used or a two-port valve in the main circuit return is also possible. A pump is required in the circuit which is the main disadvantage of this method. Some of the reduction in temperature may be achieved at the central supply point for the circuit which is termed a weather compensation circuit (see section 3.2.5).

4. Lower volume flow to heat emitter using two-port valve

   In this method, the volume flow is reduced by means of a modulating two-port valve with the flow temperature maintained. The return temperature from the heat emitter falls as the load decreases and both volume flow and return temperatures are lower in the main heating circuit. As an example, a TRV is designed to modulate the flow in response to room temperature, reducing the flow rate as the required load decreases.

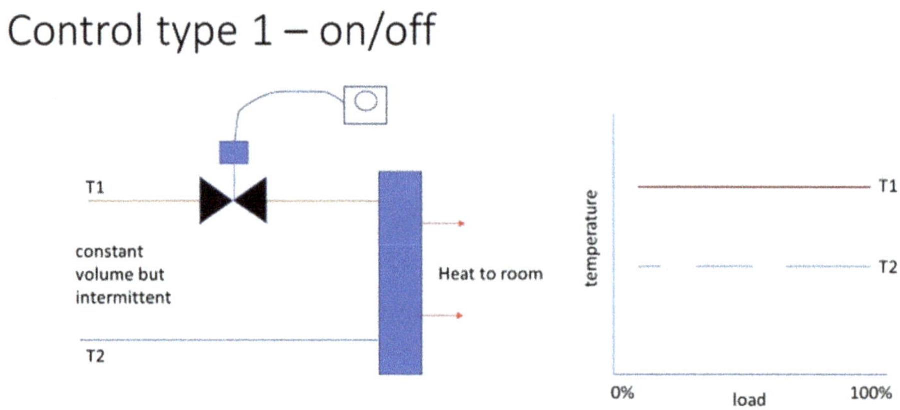

Figure 3.4. Control method 1—on/off control.

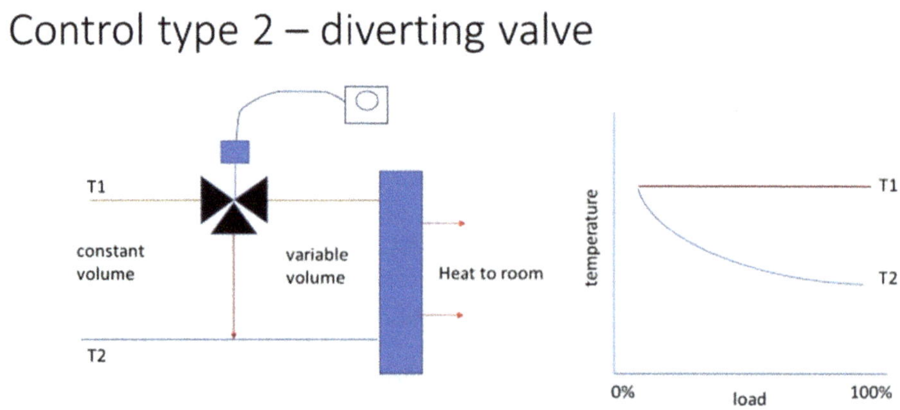

Figure 3.5. Control method 2—diverting valve control.

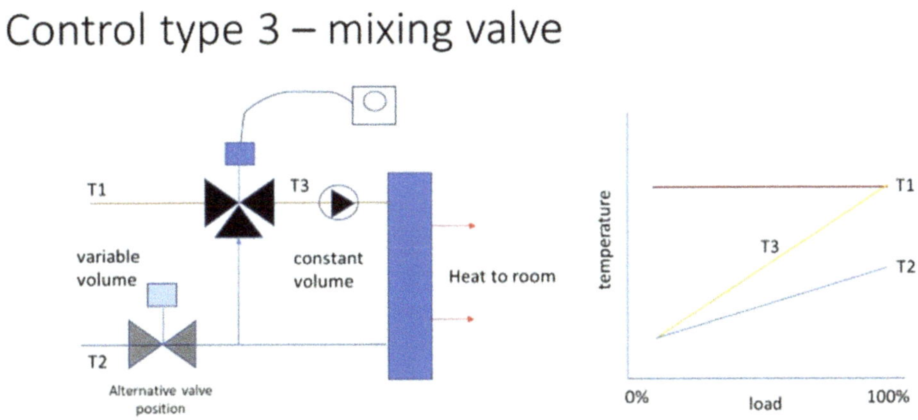

Figure 3.6. Control method 3—mixing valve control.

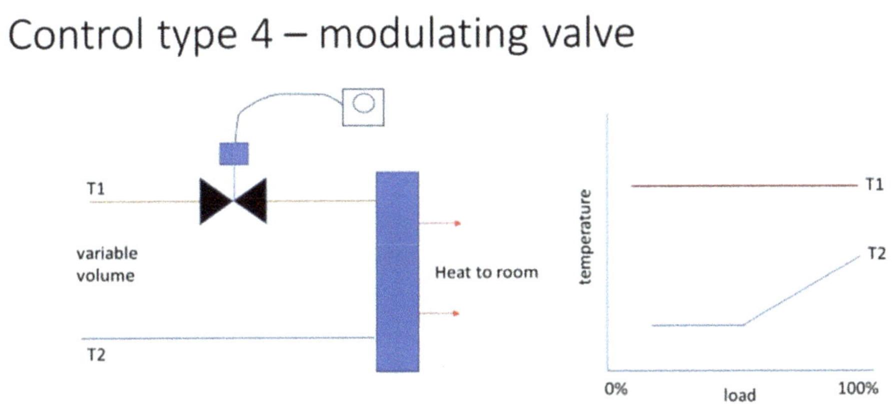

Figure 3.7. Control method 4—modulating control.

Of these control methods, it can be seen that method 2 is the least compatible with DH as the return temperature is higher than the other methods. Method 4 is the most advantageous for DH as the return temperature is kept low and a pump is not required. However, method 3 does not require a constant flow temperature and this would, therefore, be suitable for heat pump systems. Low return temperatures in the heating system will enable the DH network to operate with low return temperatures which will lead to lower network heat losses and lower pumping energy.

Although both control methods 3 and 4 have a falling return temperature characteristic a detailed calculation shows that with method 4 the reduction is faster, see figure 3.8 which plots the return temperatures for a reduction in load down to 50% using equation (3.1).

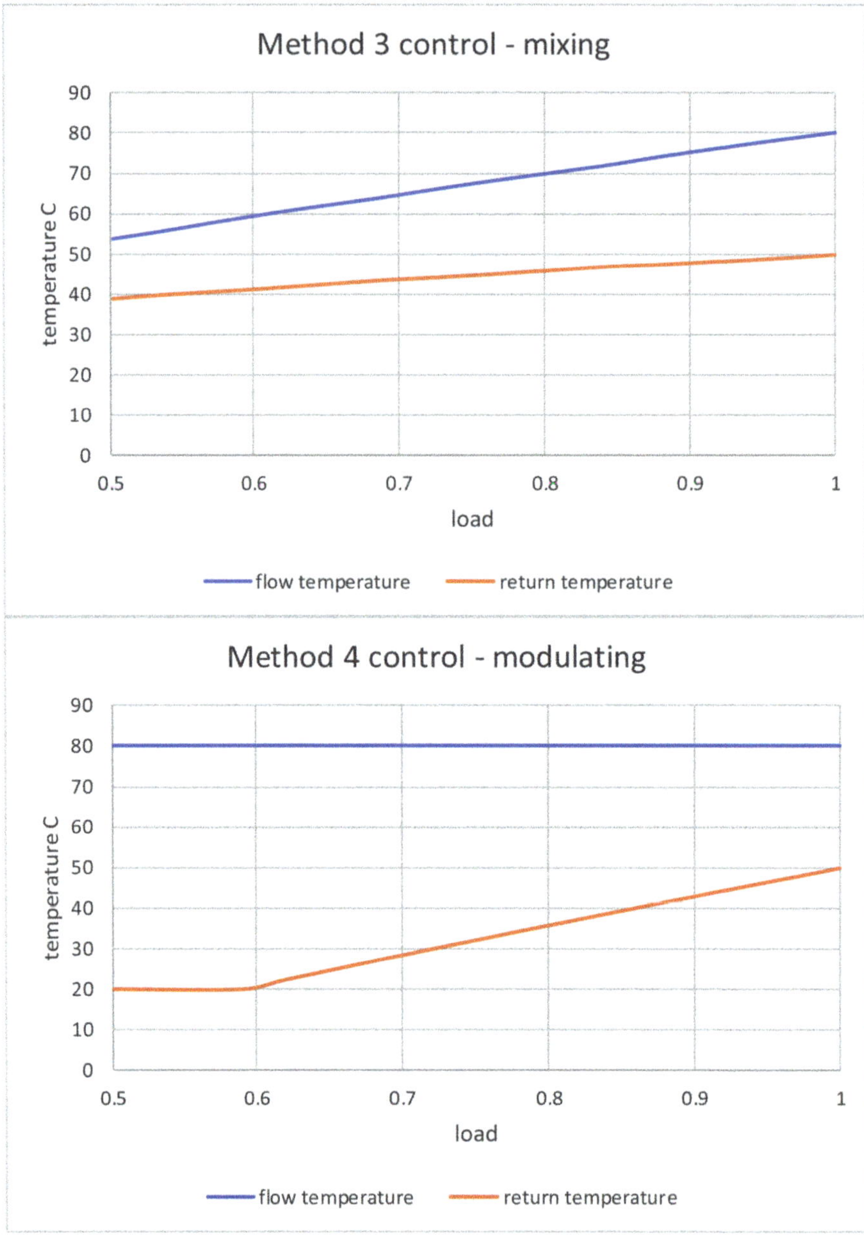

Figure 3.8. Comparison of methods 3 and 4 with an 80 °C/50 °C circuit at full load.

Connecting existing buildings

Most dwelling heating systems using radiators can be connected to DH without major modifications, the exception being one pipe systems where, if radiators are turned off, the flow in the circuit is maintained leading to high return temperatures. A wholesale redesign of the pipework to create a two-pipe flow and return system would be needed in this case.

In existing non-residential buildings there are a number of possible control systems some of which will be sub-optimal when used with DH, especially three-port diverting valves (method 2), and these will need to be modified (see section 3.7).

In conclusion, for efficient DH the control systems that result in a lower flow rate in the heat emitter circuits, i.e., TRVs for radiators and two-port control valves, are preferred as the return temperatures in the heating circuit will be lower which will enable lower return temperatures to be achieved in the heat network. This will result in lower heat losses, lower pumping energy and in some cases improved efficiency of heat production.

3.2.5 Weather compensation

Non-residential buildings often use variable flow temperature circuits to provide a basic level of space temperature control with local temperature controls providing further fine tuning of individual rooms. The flow temperature is adjusted according to the outside air temperature so that in milder weather the flow temperature is lowered in the space heating circuit. This is known as weather compensation.

For DH this can have benefits in allowing the flow temperature from the DH network to be reduced at times of part-load, which is particularly important when heat pumps are the main heat source. However, return temperatures can be higher than they would be for a two-port control system (see figure 3.8). A building heating system which reduces flow temperatures at part-load but also uses two-port control valves in each space would be of benefit to DH schemes based on heat pumps. Generally, this approach will assist in enabling the DH flow temperature to be reduced and the return temperature to be lower than design conditions at part load.

The weather compensation curve needed to match heat emitter output with space heating demands

For DH systems using heat pumps it will be important to operate with as low a flow temperature as possible. So it is necessary to understand how the flow temperature could be reduced in relation to external temperature which determines the heat demand.

For the example given below we assume that the no load condition occurs for an external temperature of 16 °C (internal gains typically provide comfortable temperatures above this) and that peak demands are set at a temperature of −3 °C. A linear relationship is assumed for intermediate heat demands. Radiator circuit temperatures at peak load are assumed to be 80 °C/50 °C.

Figure 3.9 shows the reduction in secondary heating circuit flow temperature according to external temperature such that the heat demand will still be satisfied. The calculation requires two simultaneous equations to be solved: equation (3.1) for the radiator output and equation (3.3) for the radiator return temperature.

This theoretical calculation could be supplemented by monitoring data if available. A relationship could be established between gas use and external air temperature on an hourly basis and from this the required heat in relation to external air temperature can be established.

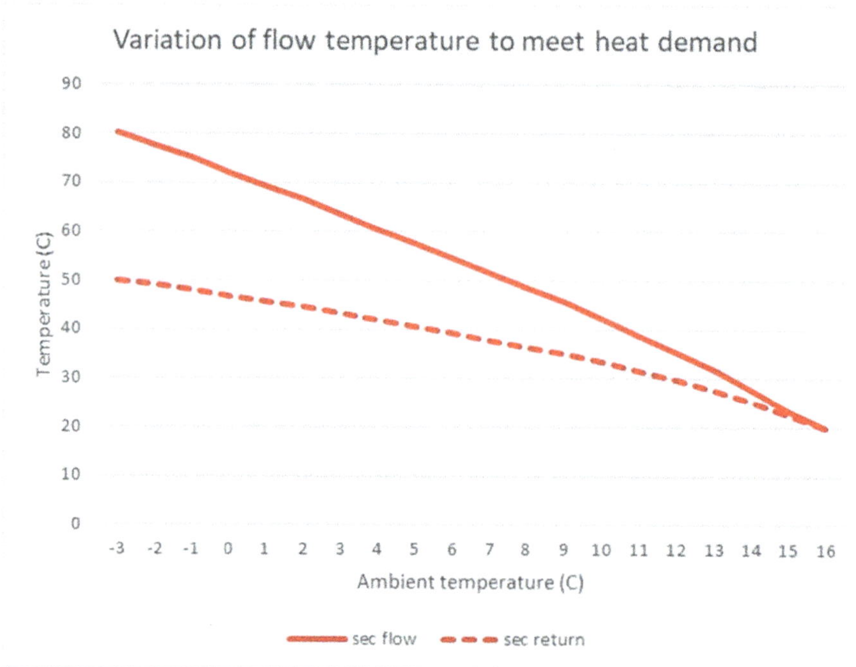

Figure 3.9. Variation in space heating circuit flow temperature to match building space heating demand.

It can be seen that there is a significant opportunity to reduce the flow temperature from the DH (albeit at the expense of the return temperature being higher than for the constant temperature variable flow control). However, in most cases it is the domestic hot water (DHW) heating requirements that will determine the minimum flow temperature of the DH (see section 3.3.3). Operating temperatures are discussed further in chapter 7.

3.3 Domestic hot water heating

3.3.1 Assessing peak demands

A good understanding of the pattern of DHW use will be of benefit when deciding on both the type of DHW heating system and its capacity. For existing buildings it may be possible to monitor hot water use as often water meters are fitted to the cold water supply to the DHW plant and many of these have pulse outputs. Logging of the pulse output will enable the actual DHW demand pattern to be established.

3.3.2 Equipment for heat transfer

Although the DHW annual demand is usually smaller than the space heating demand, the various constraints and opportunities associated with DHW heating have an important impact on the design and operation of the DH network.

There are five broad types of DHW heating available:

- Direct-fired water heaters (usually gas)
- Electric immersion heaters
- Storage cylinder or calorifier with internal heating coil
- Instantaneous hot water using plate heat exchanger (PHE)
- Storage cylinder with external PHE

Direct-fired water heaters and electric immersion heaters are not compatible with DH and will need to be changed or modified. The remaining three options can be used with a DH heat supply. These are illustrated in figures 3.10–3.12.

The cylinder or calorifier with an internal coil (figure 3.10) is commonly used both for domestic and non-domestic applications. The design with a heating coil near the bottom of the tank promotes mixing of the water. As a result, except when heating up from cold, the return water to the DH system is above the tank temperature and will typically be 50 °C–65 °C.

In dwellings, combi boilers are now the most popular choice as the hot water supply is not limited by the storage volume and the saving of the space required for a cylinder can be valuable. With DH, the equivalent to a combi boiler is to use an instantaneous plate heat exchanger (figure 3.11) which by operating in countercurrent mode results in low return temperatures of 20 °C or less as the water can be cooled to near the cold feed temperature without any risk of mixing. This type of system is commonly used both for dwellings and commercial buildings as it is cheaper and more compact.

For some buildings, DHW storage may be considered necessary where high peak demands may occur over a short time period, e.g., hotels and leisure centres. In these applications, a storage cylinder fitted with an external plate heat exchanger can be used with DH (figure 3.12). This reduces peak demands on the DH and enables the return temperature to the DH to remain low during periods of draw-off. It is important to design the tank to limit mixing and to achieve stratification with the colder water below the hot water. The circulating pump is controlled by the temperature in the tank. The hot water return is returned to the upper part of the tank which helps limit the frequency at which heat is drawn from the DH.

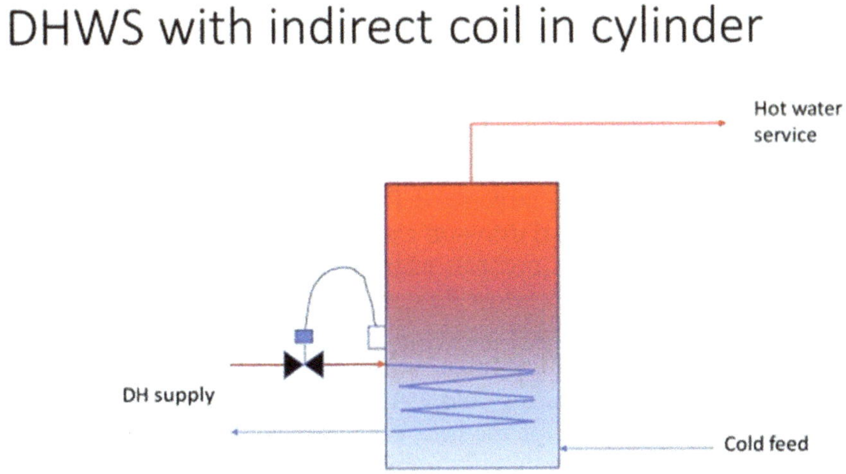

Figure 3.10. Hot water cylinder with indirect coil.

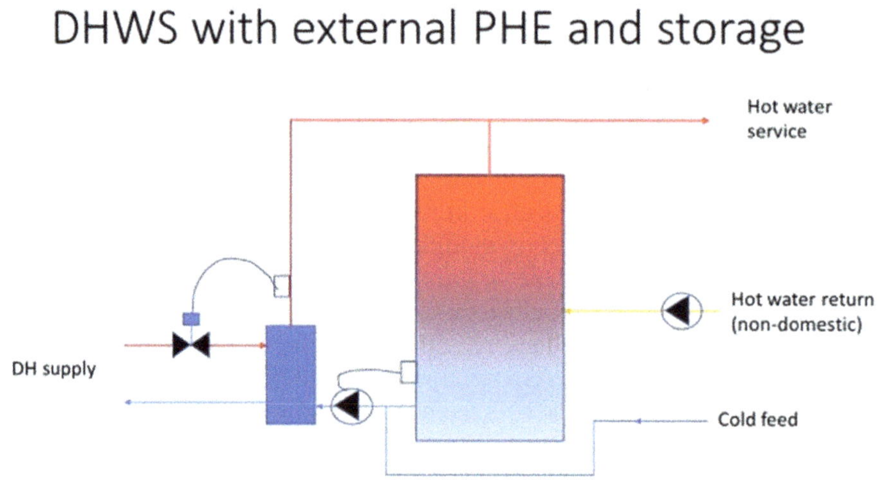

Figure 3.11. Instantaneous hot water using PHE.

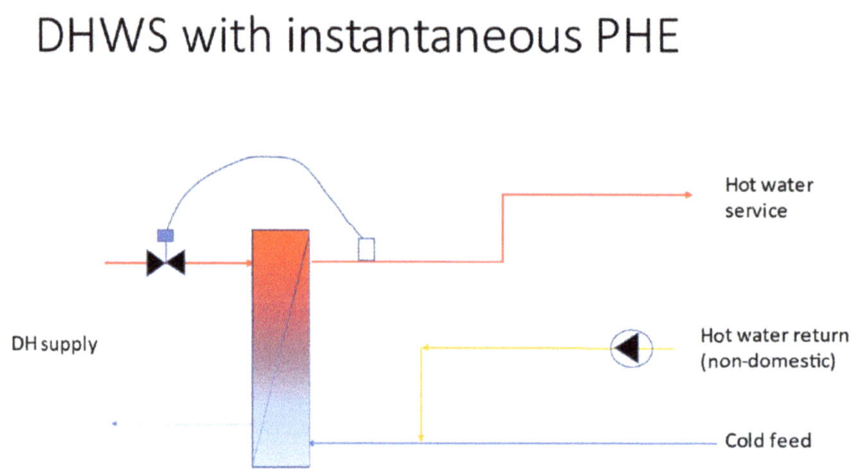

Figure 3.12. Hot water cylinder with external PHE.

The main disadvantage with large-scale DHW systems (with a DHW return) is that when there is no or minimal draw-off the required heat demand to balance the heat losses in the DHW circuit involves heating from 55 °C to 60 °C. As a result, even if it is only associated with a low flow rate, at these times return temperatures on the network will be above 55 °C.

An alternative strategy is to use trace heating for the DHW circuit. This means that the DH is only used for heating cold water and thus return temperatures can be kept low. As there is only one pipe the heat loss is halved which compensates for the electricity used which will typically have a higher price and carbon content than the DH supply.

A further option is to use local DHW heaters nearer the point of use to avoid the need for a DHW return circuit. This will mean keeping the main heating circuit live during the summer and restricting the temperature range it can use for weather compensation.

3.3.3 Operating temperatures

Instantaneous hot water can be produced at a temperature as low as 50 °C as legionella growth is unlikely with these low water volume systems [8]. It is this temperature which determines the minimum flow temperature needed from the DH, i.e., if all customers are using instantaneous systems, as on a small housing scheme, the flow temperature could be lowered to around 55 °C–60 °C.

If a storage system is used then it will need to store hot water above 60 °C to prevent legionella formation so the DH temperature will have to be a minimum of about 65 °C–70 °C even if only one building on the network uses storage. Hence, when considering the adoption of lower operating temperatures it is the customer's requirements for domestic hot water heating that governs the minimum flow temperature that can be supplied by the DH.

3.3.4 Control of DHW temperature

Non-domestic buildings

Commercially available DHW heaters using a PHE are frequently provided with three-port control valves as standard which if used with DH will result in high return temperatures. It is essential for two-port control valves to be specified as in figures 3.11 and 3.12.

Dwellings

Although the instantaneous hot water heat exchanger is commonly used for dwellings and normally incorporated as part of the heat interface unit (see section 4.4) there are some design issues that need to be considered.

The control valve that controls the DHW temperature needs to act quickly and in a stable manner when a rapid hot water draw-off occurs. Otherwise, the temperature could be slow to reach its set point or overshoot and risk causing scalding.

There is a need to keep the heat exchanger (or the local primary pipework) warm so that on hot water draw-off the response time is satisfactory; CIBSE Code of Practice CP1.2 [9] requires the temperature at the tap to be 45 °C after 45 seconds. There are two approaches:

- By having a standby or 'keep warm' setting. A small flow is introduced to maintain the heat exchanger at around 40 °C. This may be a continuous flow or intermittent pulses. In the standby condition a small flow rate with a return temperature of around 45 °C is needed which means the flow pipe will also be kept slightly above this temperature at all times.
- By having very short branches from a riser to the dwelling and having temperature-controlled bypasses in the riser.

Both of these approaches have been used successfully and are permitted under the BESA testing regime for HIUs (see section 4.3.2).

3.3.5 Comparison of systems

At the building level

Instantaneous heat exchangers have the benefits of compactness and low cost compared to the storage solution which involves more space and costs especially when designed to deliver mains pressure hot water. This cost advantage is important when comparing DH with individual heat pumps for dwellings as a storage cylinder is necessary for the latter.

Storage of hot water involves heat losses even with well insulated cylinders whereas heat losses from insulated instantaneous heat exchangers are much lower. The cylinder heat losses in dwellings can be considered useful heat in winter but would lead to unwanted heat gains in summer and a risk of overheating.

Impact on the DH network

An instantaneous DHW system will result in peak demands coincident with the demand for DHW. The peak demand on the DH network will depend on the diversity factor applied—see Annex D in CIBSE CP1.2 [9]. In contrast, a storage system will have a lower peak demand requirement at each dwelling depending on the selection of the heat-up time and the time that this peak demand occurs can also be varied by use of a programmer. The difference in flow rates is mitigated somewhat as the instantaneous system will achieve lower return temperatures and hence a lower flow rate for the same level of demand.

As the use of the instantaneous heat exchanger will be for a short period some designs shut off the space heating circuit to prioritise the supply to the hot water heat exchanger. This reduces peak demands marginally.

When diversity is applied the total flow rate (space and water heating) required for an instantaneous DHW system is typically comparable to a system with hot water storage after more than about 20 dwellings are connected.

A storage system could result in lower network heat losses by using intermittent hot water heating. For example, the hot water store could be heated only at night. This would mean that the DH supply to that customer would be shut down for much of the day in summer. The local DH pipework would then cool down in the off period which would reduce losses from both the flow and return pipework. Given that there is storage the longer response time when the system is reactivated would not be of concern. Hence storage systems and intermittent heating could be advantageous where there are long runs of branch pipework to individual buildings where the heat losses associated with the standby condition would otherwise be high. The impact of frequent cycling of temperatures on the integrity of the branch pipe will need to be evaluated (see section 5.7).

A comparison of the two types of DHW systems is given in table 3.3 below, see also Annex H of CIBSE CP1.2 [9] and the guidance in [10] and [11].

Table 3.3. Comparisons of DHWS systems.

Issue	Instantaneous	Storage
Service constraints	Unlimited supply	Limited: depends on storage volume and heat-up rate
Cost	Low	High
Space	Compact	Significant
Return temperatures	Low	Higher unless external plate heat exchanger used with stratified tank and 2-port valve
DH flow temperatures	Can be reduced to 55 °C	Needs to be >65 °C
Pipe sizing	Will be larger for local branches, but will be similar on main branches after diversity applied	Can be smaller especially when long heat-up times are used
Heat losses from DHW demand	Low from heating equipment	Unwanted summer losses from store and risk of overheating
Heat losses from DHW circuit	Can be reduced using trace heating	Can be reduced using trace heating
Heat losses from local DH network	Flow and return pipes maintained close to 'keep warm temperature'	Can be lower in final branches if intermittent heating operation used (benefit only in summer)
Standby provision	None	Storage allows for short interruptions in DH service without impact on customers (in summer). Electric immersion heater can be fitted for additional back-up

3.4 Space cooling

3.4.1 Heat transfer equipment

Fan-coil units are commonly used in commercial buildings and AHUs fitted with cooling coils. Chilled beams use higher temperatures (to avoid risk of condensation on their surfaces). This is an advantage for DC as return temperatures are higher.

Split system air-conditioning units where the refrigerant transfers coolth from the air stream directly would not be compatible with DC.

3.4.2 Operating temperatures

Most cooling systems in buildings are designed for operating temperatures of 6 °C flow 12 °C return. As a small temperature difference leads to large volume flows and so larger pipes, DC systems often aim to increase the temperature difference by lowering the flow temperature and raising the return temperature.

Lowering the flow temperature is possible, however, the efficiency of the central chillers will be reduced and temperatures below 3 °C may risk freezing in transient conditions. Raising the return temperature is possible, however, if the cooling is also required for dehumidification, then the amount this can be raised depends on the temperature the air needs to be cooled to before reheating.

Chilled beams need to operate with elevated chilled water temperatures to avoid condensation on surfaces and typically 14 °C/17 °C circuits are used. Hence a good solution for DC would be to use say 3 °C/14 °C for the main chilled water circuit with chilled beams being supplied in series from the return of this circuit.

If there is a heat exchanger between the DC network and the building's cooling circuit then an additional temperature drop of 1 °C to 3 °C will be required. This may mean that lower temperatures are needed from the chillers impacting efficiency or that the ΔT in the DC network has to be smaller. A few degrees difference in a DH network operating with a ΔT of 30 °C would not be too critical. However, for a DC circuit operating with a ΔT of only around 10 °C every degree difference will have a significant impact on flow rates and pipe sizing. Direct connection of buildings without using a heat exchanger is, therefore, preferable.

3.4.3 Control of space temperatures

As for the control of space heating (see section 3.2.3), for cooling systems the use of two-port controls and variable volume is preferred so that the return temperature is kept high and pumping energy low. The pumping energy is more important for DC as the temperature rise due to friction in the pipework adds to the cooling energy required whereas in DH this energy is transformed into additional useful heat.

3.5 Modifications to existing building heating systems

Connection to existing buildings can usually be achieved without changing the existing heating system and heat emitters (steam heater batteries being an example where wholesale changes would be needed), however, the existing heating system will have an impact on the design and operation of the DH/C system which may then operate sub-optimally for a long time. It may be that just one building influences the design of the system disproportionately, e.g., if a building has been designed for a high flow temperature at peak this would determine the flow temperature of the entire network.

It is, therefore, sensible to consider modifications to the customer's existing building heating systems to allow the DH operator to optimise operation by varying DH flow temperatures and to minimise the costs of connection of the building to DH.

The following modifications are considered the most useful and worth developing with the customer. Apart from the last one, they do not involve significant disruption to the customer.

Establishing whether heat emitters are oversized in relation to the peak demand requirement

This can be done in three ways:

- Pogressively reducing the flow temperature below the current set point and monitoring internal space temperatures.
- Using the BEMS to monitor the position of AHU control valves—if these are still partially closed under design conditions there is spare capacity in the heating coils.
- Measuring the peak demands using heat meters and comparing this with the installed heat emitter capacity.

Rebalancing of heating circuits

Setting lower flow rates at each heat emitter will result in a small reduction in heat output but lower return temperatures in the heating circuit. Provided there is sufficient oversizing present this is an effective way to reduce return temperatures on the heat network, giving benefits in pipe sizing, heat losses and pumping energy. For example, a system set up for 80 °C/60 °C could be rebalanced for 80 °C/50 °C, by reducing the flow rate to 67% of its original design with heat output reduced by 13% (see table 3.1). One way of doing this would be to install new pre-settable pressure-independent TRVs on all radiators, potentially improving local temperature control as an additional benefit. Otherwise, existing valves may be available for balancing the circuits.

Eliminating unnecessary bypass flows

Often bypasses are included in circuits to ensure that heating zones respond quickly and to make sure there is a minimum flow in all circuits to prevent build-up of debris. For the building space heating system this approach has some disadvantages in higher pumping energy. Bypasses lead to higher return temperatures for the building's heating circuit and hence also for the DH circuit. Hence, eliminating or reducing bypass flows to a minimum by introducing additional control valves or temperature-controlled bypasses will enable return temperatures to be kept as low as possible.

Change constant volume control to variable volume control

If a system is controlled using diverting three-port control valves then under part-load conditions the return temperatures will rise. This will result in higher return temperatures in the DH network. Changing to a variable volume system by replacing or modifying control valves to operate as a two-port control system will reduce return temperatures at part-load benefiting the DH system with lower heat losses and pumping energy

Change to variable speed pumps

If a variable volume control system is introduced or if the flow is reduced at the heat emitters, existing fixed speed pumps will become less efficient and will produce more heat than is needed. Retrofitting variable speed drives, or installing new variable speed driven pumps, will enable the pump to operate at lower speed to match the lower volumes and heat required. This will reduce electricity consumption and bring a direct benefit to the customer.

Introduce weather compensation
Introducing weather compensation where this is not used could have some benefits in enabling a customer to limit the maximum output of heating systems centrally. This can be achieved as part of the function of the control valve at the DH interface connection (see chapter 4).

Reduce flow temperature at peak
As many systems will be oversized but are operated with a variable flow temperature it will often be possible to reset the weather compensation curve to reduce the peak flow temperature. This has the benefit of enabling the DH flow temperature to be lowered which may result in higher heat production efficiency.

Change to domestic hot water systems to enable lower DH flow temperatures
In most cases, the DH system will benefit from operating at lower temperatures under part-load conditions, especially in summer. However, it is the DHW design that will determine the minimum required temperature of the DH supply. If there is sufficient oversizing in the DHW equipment then lower temperatures can simply be agreed to be acceptable. If not, then the DHW heating equipment will need to be changed to maintain its capacity, e.g., by adding a new external PHE to work in conjunction with an existing PHE or storage calorifier.

Change internal heating coils to external PHE on hot water calorifiers
Where calorifiers are fitted with internal heating coils a modification can be made to retrofit an external plate heat exchanger. This can be sized to produce much lower return temperatures than usually found with conventional calorifiers. Flow rates from the PHE into the DHW calorifier need to be low and diffusers or sparge pipes fitted so that stratification is achieved as this will enable the lowest DH return temperatures to be achieved. This arrangement will result in lower return temperatures on the DH heat network reducing heat losses and pumping energy.

Reducing peak demands at start-up
Many commercial buildings will have high initial peaks when the building is heated up from cold after a night setback. This is not a problem for conventional gas boilers but artificially high peak demands are unhelpful for DH and may mean that a secondary higher carbon plant is brought on line for a short time. The control system could be modified either to ramp up the heating system flow rates or to increase the heating set-point gradually to limit peaks, even if this means starting the heating system a little earlier in the morning.

Increase heat emitter size or reduce heat losses
If there is limited oversizing, then it may be necessary to increase the heat emitter size to enable lower mean temperatures to be used to benefit the DH network and production plant. This might be achieved by adding additional radiators or changing single panel radiators to double panels. It may be necessary only to modify some of the radiators. There is less likely to be oversizing present with AHU coils as these will be engineered more closely to the design air flow rate and it is possible that AHUs will need to be modified to increase coil area.

As an alternative, investment in improving the building fabric, e.g., through adding insulation would enable lower operating temperatures to be used but avoid the cost of changing heat emitters.

Bibliography

[1] Building Regulations 2010 *Building Regulations and Approved Documents Index* The Merged Approved Documents, Approved Document L1, Clause 5.10, Page 36, HM Government March 2023 (https://gov.uk/guidance/building-regulations-and-approved-documents-index#-approved-documents)

[2] Efficient Substations and Installations 1996 *Report N5* (IEA District Heating and Cooling Annex) P 58

[3] CIBSE 2021 *Domestic Heating Design Guide* 10th edn (CIBSE)

[4] Department for Business, Energy and Industrial Strategy 2021 *Domestic Heat Distribution Systems: Evidence Gathering, Final Report*, BEIS Research Paper Number: 2021/015 (https://assets.publishing.service.gov.uk/government/uploads/system/uploads/attachment_data/file/976021/beis-dhds-final-report__1_.pdf)

[5] Oventrop *Thermostatic Radiator Valves Technical Information* (https://oventrop.com/en-GB/productssystems/articledetails/1180604)

[6] Danfoss *Data Sheet Dynamic Valve Type RA-DV Pressure Independent Radiator Valve* (https://assets.danfoss.com/documents/207391/AI125086402935en-010603.pdf)

[7] Rogers and Mayhew 1992 *Engineering Thermodynamics Work and Heat Transfer* 4th edn (Longman Group) p 620

[8] CIBSE 2021 *Guidance Note: Domestic Hot Water Temperatures from Instantaneous Heat Interface Units* (CIBSE)

[9] CIBSE 2020 *Heat Networks: Code of Practice for the UK, CP1.2* (CIBSE)

[10] Parsloe 2021 *Chris Design Guide: Heat Networks, Section 6.2.1* (CIBSE)

[11] Frederiksen S and Werner S 2013 *District Heating and Cooling* (Studentlitteratur) P 374

Chapter 4

Interface between building heating and cooling systems and the DH/C systems

4.1 Principal requirements for DH interface

The DH system supplies heat to the building through interface equipment normally referred to as a heat or thermal substation or in the case of dwellings a heat interface unit (HIU).

The primary requirements of the interface are to provide:

- A means of isolation of the building heating system and the DH system in the event of a fault
- Control of the maximum DH flow rate at the building so that the flow in the DH network is balanced and all customers receive their required flow rate
- A point at which the heat can be metered

Secondary requirements of the interface equipment typically include:

- Control of the flow temperature in the building's heating system
- Control of the supply temperature of the domestic hot water service
- Heat exchanger to separate the DH system water from the building heating systems (if required—see section 4.2)
- Secondary pump for the building heating system (if required)
- Strainers to protect interface equipment from debris either in the building's heating system or the DH system
- Instrumentation and sensors to indicate the pressures and temperatures in each part of the circuits
- A bypass circuit or a control system that keeps the interface equipment up to temperature during times of no load so that adequate response times are achieved

These functions are discussed further in sections 4.3 and 4.4.

doi:10.1088/978-0-7503-5286-4ch4

4.2 Types of interface equipment

4.2.1 Methods of connection

There are two ways in which a building or dwelling can be connected to a DH/C network, termed direct connection and indirect connection.

- A direct connection is where the water in the DH/C network flows directly into the heat emitter circuit or chilled water circuit within the building.
- An indirect connection is where a heat exchanger is placed between the DH/C network and the building's heating and cooling systems creating two separate circuits.

The choice of connection and location of hydraulic breaks within the circuit is an important strategic design decision which will need to be taken early in the scheme development as it will be of concern to potential customers. The strategy will influence the design operating temperatures for the network, the capital costs, operational efficiency and the responsibilities of the customer and DH/C operator.

A direct connection system will benefit the DH/C system by eliminating the temperature difference between the DH/C network and the building's systems. This means that the DH circuit can operate at a lower temperature and the DC circuit at a higher temperature resulting in potential benefits in lower heat/coolth production costs and lower heat losses/gains. This is particularly important for heat pump systems where higher flow and return temperatures will impact the CoP (see section 6.3).

For a direct connection to be possible the pressures in the DH/C system and the building's heating system need to be compatible and concerns over water treatment and responsibilities for water quality and leaks need to be resolved and agreed between the DH/C company and its customers.

4.2.2 Pressure constraints

For a direct connection, the maximum expected pressure on the DH/C network will need to be less than the pressure rating of the building's heating system at all times, including under certain fault conditions.

If a tall building is to be connected to the DH/C, the static pressure required by the tall building would increase the pressure in the DH/C network unnecessarily and an indirect connection would be preferred.

A DH/C network developed in an area with a large variation in ground levels may need a relatively high static pressure limiting the scope for direct connections.

Generally, it is found that indirect connection is appropriate for large non-residential buildings as the additional cost for the indirect connection is less significant. For dwellings, direct connection is preferred if the pressure ratings are suitable and risk of damage from leaks is acceptable as it is a simpler, lower cost design.

When dealing with existing buildings it may be unclear what the effective pressure limits are and to avoid undue risk the customer may not accept direct connection at a higher pressure than currently used and so an indirect connection may be necessary.

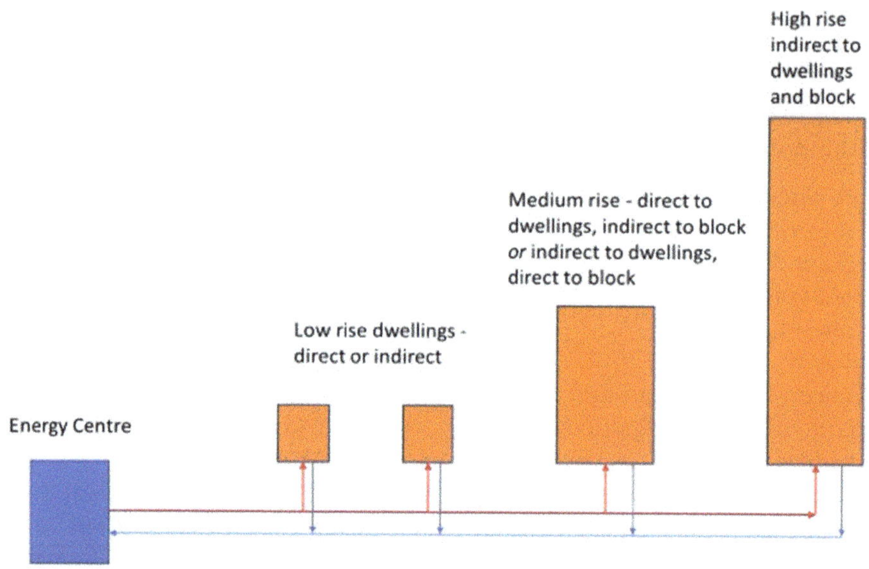

Figure 4.1. Potential options for methods of connection.

Even where indirect connection of dwellings is used the primary side pipework supplying the HIU may have much higher pressures than commonly seen in dwellings, especially for very tall blocks, but will still have to be routed within the dwelling. Some additional safety measures may need to be considered.

Figure 4.1 illustrates some of the typical situations and solutions.

For low rise dwellings the cost of the network connection is likely to be relatively high and the risk of damage to other properties is limited, so minimising cost through direct connection would be preferred.

For a medium rise block of flats the increase in network static pressure associated with connecting the block is not normally sufficient to require indirect connection. So the block could either be directly connected with indirect connection of dwellings if risk of damage is of concern or the block could be indirectly connected with direct connection of each dwelling if cost is a major concern. Indirect connection of both the block and the dwellings would add to costs unnecessarily and increase network return temperatures.

For the high-rise block it will be necessary to have an indirect connection for the block to avoid increasing network static pressure and the pressure within the block may mean that indirect connection is necessary at the dwellings as well (or at least for the lower floors).

4.2.3 Water quality and leaks

Whilst the DH/C system water may be treated to a high level of quality and so suitable for supply to building heating systems with direct connection there is a risk to the DH/C system from contamination from the circuits within the building, especially where an older existing building is to be connected. Also, in some cases the water treatment used in the DH may not be compatible with the materials used within the building's systems.

Direct connection means that any leaks from within the building will need to be replaced by adding water to the system by the DH/C company but it would be difficult for the DH/C company to take action to remedy such faults.

If there is a serious leak within a property, then there would effectively be unlimited quantities of water available to keep the leak flowing. This has been of particular concern to owners of high-rise dwellings where damage could occur both to the property where the leak occurs and others on lower floors. The provision of easily accessible and operable isolation valves will reduce the risk but some properties may be unoccupied for long periods.

4.2.4 Typical arrangements

The minimum requirements for building interfaces are given in Objective 3.3 in CIBSE CP1.2 [1] together with a discussion on the merits of direct and indirect connection.

Schematic arrangements for substations/HIUs are shown in figure 4.2 for indirect connection, figure 4.3 for direct connection and figure 4.4 for direct connection with flow temperature control.

The primary requirement for limiting flow rate to a maximum design flow is shown in the diagrams by means of a differential pressure control valve (DPCV). This can be adjusted to set a pressure difference that is independent of the pressure difference locally in the DH network. This pressure difference then determines the maximum flow through the circuit. In some cases, an additional balancing valve may be needed. In figures 4.2, 4.3 and 4.4 a separate DPCV and two-port control

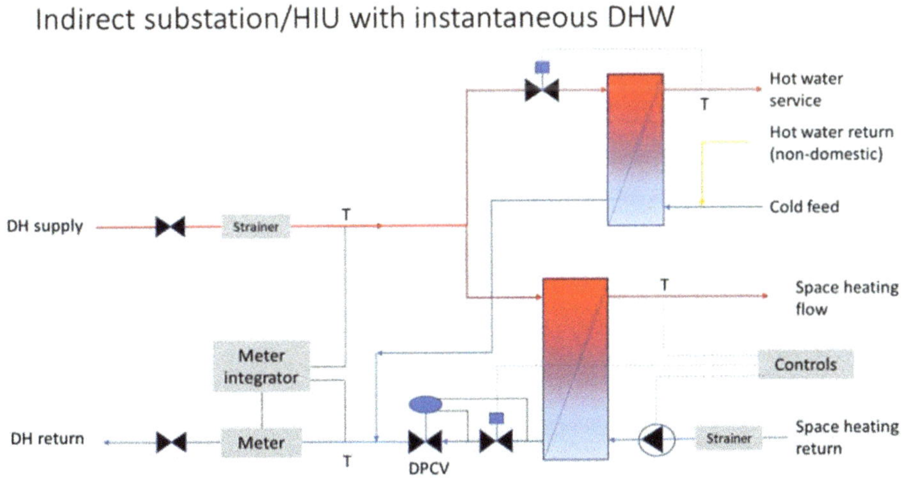

Figure 4.2. Indirect substation schematic (simplified).

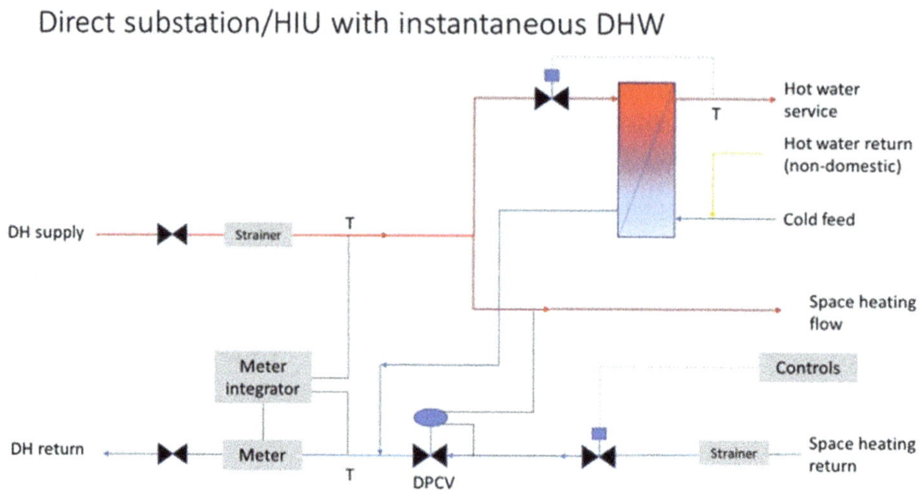

Figure 4.3. Direct substation schematic (simplified).

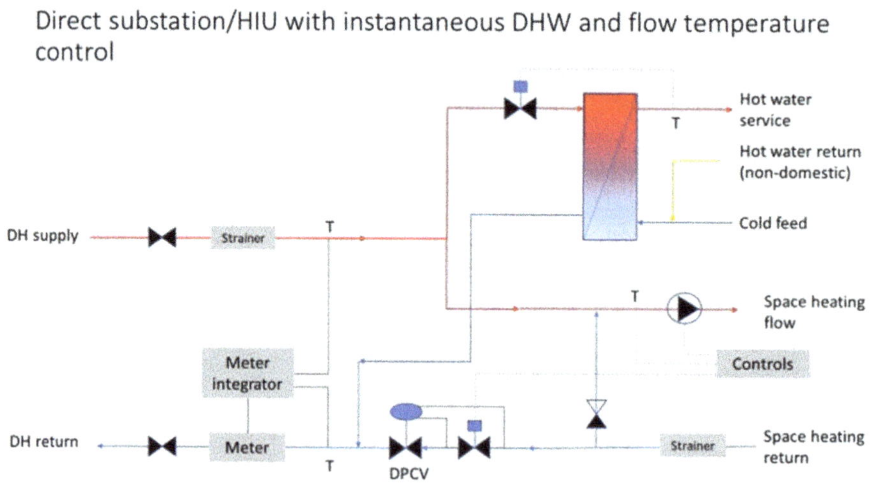

Figure 4.4. Direct substation with flow temperature control schematic (simplified).

valve is shown. It is possible to combine these two functions in one valve termed a pressure independent control valve (PICV).

4.3 Heat interface units for dwellings

4.3.1 Principal components

Since 1990 there has been a trend towards packaging the necessary controls and heat meter in a compact, factory-assembled unit known now as a heat interface unit (HIU). Often these units include a domestic hot water plate heat exchanger to produce instantaneous hot water. They may be designed for a direct connection or an indirect connection, the latter including a second plate heat exchanger, temperature control valve and pump for the space heating circuit. The schematic diagrams in figures 4.2 and 4.3 illustrate the principal components and these are described below.

Isolation—valves are required to aid installation and commissioning but also available to the occupant for emergency shut-off in the event of a leak. Quarter turn lever operated ball valves are most suitable.

Strainer—this is needed to protect the equipment from debris on both the primary and secondary sides.

Heat meter—this will measure the heat taken by the property (in kWh) and will also be able to provide useful information on volume flow and flow and return temperatures which can be used to identify faults, e.g., abnormally high return temperatures. An automatic meter reading is often used to enable remote reading of heat meters and other operating data (see section 8.3).

Domestic hot water (DHW) heat exchanger and control valve—this will typically be sized to deliver hot water at 50 °C–55 °C [2]. Primary return temperatures are selected at 20 °C or lower. An allowance should be made for scaling over time. The control valve needs to be fast-acting to avoid high DHW temperatures that could cause scalding and operate over a range of system pressure differentials and a wide range of DHW draw-off rates. Some designs shut off the space heating flow when DHW is required. As this part of the HIU is connected to the mains water supply WRAS approval is needed.

Standby flow/keep warm facility (if applicable)—to make sure the response time (the time that hot water arrives at the tap after turning on) is sufficiently short the heat exchanger may need to be kept warm during periods when there is no water use. The flow rate to do this needs to be well-controlled otherwise high return temperatures will result. This is important as typically the HIU will be in standby mode for 85% of the time.

Space heating heat exchanger and control valve (for indirect connection)—this will be sized to deliver the required space heating circuit heat rate and operating flow temperature with approach temperatures on the return typically less than 3 °C. The primary side control valve will be set to control the secondary side temperature at a constant or variable temperature. This valve needs to be shut off when no heating is required to avoid bypassing. The primary side control valve should either be a PICV or a separate DPCV should be included so that when the control valve is fully open the flow can be limited to the set flow rate for the dwelling. This ensures the system is in balance regardless of the pressures in the network.

Space heating circulation flow—for an indirect connection and variable temperature direct connection, the secondary side space heating circuit will require a pump to provide the necessary circulation pressure. This needs to have an appropriate head for the circuit and the control valves such as TRVs in the circuit need to be able to close against this head. The pump will be linked to the dwelling programmer and room thermostat and turned off when not required. An indirect space heating circuit needs to include an expansion vessel.

On/off control—the HIU will be linked to a programmer able to schedule on/off times for the space heating circuit and with a simple on/off button to override the programme.

Pre-payment systems—these may be included and will involve a two-way communication with the HIU to open a primary side on/off valve when the account is in credit and shut off when payment is overdue.

Instrumentation—although the heat meter display will indicate the primary side temperatures, other temperature gauges and pressure gauges may be incorporated. It is helpful for troubleshooting to have pressure gauges or pressure tappings available so that the primary pressure differences available (upstream and downstream of the DPCV) can be established.

Packaging and insulation—the HIU equipment is normally contained in a casing with an attractive finish and insulation may be applied to individual components, or to the casing as a whole, to limit heat losses from the unit to less than 1 kWh per day as required by [1].

There are a number of suppliers of HIUs with differing designs and approaches. Both direct acting mechanical controls and electronic controls are used.

4.3.2 HIU testing

Experience has shown that the design, commissioning and ongoing monitoring of HIUs has a significant impact on the efficiency of a DH network. To improve the performance of HIUs and provide a standardized set of information that designers can use, a set of tests has been developed by the Building Engineering Services Association (BESA) [3]. HIUs that have been tested and passed are provided with a BESA certificate of registration and the results of the tests are published on the BESA website [4].

The scope of the testing covers: safety in relation to risk of scalding, customer service levels, primary return temperatures and overall energy efficiency. The test data enables the Volume Weighted Average Return Temperature (VWART) to be calculated for each circuit and mode of operation separately and in combination. The combined VWART gives an indication of the average return temperature on the network that would be expected under typical operating conditions (see below).

The tests are modular which allow different types of HIU to be included in the regime. The test modules and test temperatures are defined in table 4.1.

Modules 9 and 10 apply to HIU designs which do not include a keep warm function. These can be used where there are controlled bypasses in the local network and short branches to the HIU (see section 4.5).

The tests aim to simulate operating conditions over a range of typical values:

- Primary differential pressure is set at 50kPa and 200kPa
- Space heating demands are set at 0.5kW, 1kW and 4kW
- DHW draw-off rates are varied from 0.02l/s to 0.13l/s (with higher rates used to determine maximum rate of hot water delivery)
- Cold water supply is kept at 10°C

Table 4.1. Test modules for HIUs in the BESA test regime

Module No.	Heating circuit/type	Primary flow temperature (°C)	Secondary temperatures (°C)
1	Space heating-indirect	70	55/35
2	Space heating-indirect	55	45/35
3	Space heating-direct	70	70/35
4	Space heating-direct	55	55/35
5	Space heating–direct mixing	70	55/35
6	Space heating–direct mixing	55	45/35
7	DHW with keep warm	70	50
8	DHW with keep warm	55	50
9	DHW without keep warm	70	50
10	DHW without keep warm	55	50

Table 4.2. Pass criteria for the BESA HIU tests–space heating

Module	VWART (°C)	Space heating differential pressure (kPa)	Space heating temperature (°C)
1	<40	-	-
2	<40	-	-
3	-	>10 <60	-
4	-	>10 <60	-
5	-	>10 <60	55 +/– 5
6	-	>10 <60	45 +/– 5

The testing includes pass/fail tests as well as best practice levels for certain parameters. Results are presented graphically to show the dynamic response under simulated operating patterns. Tables 4.2 and 4.3 detail the main criteria for units to pass.

The primary return temperature criterion is related to the need to control scale formation within the heat exchanger. There are also criteria for the stability of the DHW flow temperature during initial draw-off periods-it must not exceed 60°C for more than 1 second to reduce risk of scalding and it must not drop below 45 seconds for more than 5 seconds. The average DHW temperature must be maintained at 50°C +/– 3°C during the initial draw-off periods and in the low-flow rate test at 0.02l/s.

Table 4.3. Pass criteria for the BESA HIU tests – DHW heating, includes:

Module	VWART (°C)	VWART (no load) (°C)	Steady state DHW temperature (°C)	Primary return temperature (°C)	Time to reach 45°C s
7	<22	<44	50 +/− 1	<55	<15
8	<27	<48	50 +/− 1	<55	<15
9	<22	<40	50 +/− 1	<55	<15
10	<27	<40	50 +/− 1	<55	<15

In addition to the above, the energy losses for the HIU including heat losses and electrical use shall be <1kWh per day, based on measurements over 8 hours in the standby (no-load) condition.

Calculations of volume-weighted average return temperature (VWART)
The tests above provide primary return temperatures for space heating, hot water draw-off, keep warm and non-keep warm conditions. These are weighted by a standardised calculation in Appendix C of [3] taking account of volumes and operating times for each operating condition to provide the following VWART figures:

- Volume weighted average return temperature for hot water draw-off ($VWART_{HW}$)
- Volume weighted average return temperature for space heating ($VWART_{SH}$)
- Volume weighted average return temperature for keep warm ($VWART_{KM}$)
- Volume weighted average return temperature for non-keep warm ($VWART_{NKM}$)

These figures are used to calculate:

- Volume weighted average return temperature for the winter period when space heating is being used ($VWART_{WINTER}$)
- Volume weighted average return temperature for the summer period when there is no space heating ($VWART_{SUMMER}$)
- Volume weighted average return temperature for the year as a whole ($VWART_{HIU}$)

These VWART figures will assist the designer in comparing HIUs and estimating heat losses from the network. They also provide a useful benchmark against which to compare operating data and hence identify HIUs that have a malfunction.

4.4 Substations for non-residential buildings

A schematic for a typical indirect substation is given in figure 4.2. It is preferable to have a separate DHW heat exchanger rather than having this supplied from the secondary heating circuit as the minimum DHW loads can be much smaller.

Although a single plate heat exchanger is shown for the space heating circuit it is preferable to have two heat exchangers so that the wide variation of load can be followed with the heat exchangers operating in duty/assist mode. This improves control, avoids the change to laminar flow at very low demands (which reduces heat transfer rates) and results in a more stable return temperature. It also provides some redundancy should there be a failure in the heat exchanger or control valve.

Plate heat exchangers designed for counterflow operation can achieve approach temperatures (the difference between primary return and secondary return) of 5 °C or less.

The equations for the design of the heat exchanger are given in section 3.2.3. These can be used to determine the increase in surface area required to reduce the approach temperature from 5 °C to 3 °C. For example, suppose a heat exchanger is sized initially to deliver Q from a primary circuit operating at 85 °C/55 °C and a secondary circuit at 75 °C/50 °C. To reduce the primary return to 53 °C will change the LMTD so from equation (3.1):

$$Q = UA_1\, LMTD_1 = UA_2\, LMTD_2$$

$$LMTD_1 = [(85 - 75) - (55 - 50)]/ \ln[(85 - 75)/(55 - 50)] = 7.21\ ^\circ\mathrm{C}$$

$$LMTD_2 = [(85 - 75) - (53 - 50)]/ \ln[(85 - 75)/(53 - 50)] = 5.81\ ^\circ\mathrm{C}$$

Hence $A_2/A_1 = LMTD_1\ /LMTD_2 = 7.21/5.81 = 1.24$, i.e., a 24% increase in surface area of the heat exchanger is required to lower the return temperature by 2 °C.

The same calculation for reducing the approach temperature from 5 °C to 2 °C results in a 45% increase in the surface area of the heat exchanger.

The maximum primary flow rate is limited by the DPCV (or PICV) which is connected to maintain a set pressure difference across the control valve regardless of the pressure difference in the network. This limits the maximum flow that can be taken and provides good control conditions for this valve. This control valve then responds to load by controlling the secondary flow temperature resulting in variable volume flow in the primary DH circuit and low return temperatures.

Where the customer has agreed to variable secondary flow temperatures (weather compensation) the primary control valve can be used with a variable set point according to the customer agreement. It is important for the control logic to default to a set point a few degrees lower than the flow temperature of the DH should the DH supply temperature be below that required. Otherwise, the control valve will become fully open for no benefit leading to high return temperatures.

Where there are significant DHW demands a more complex substation may be used, termed a two-stage substation. The DHW is heated in series, firstly by a heat exchanger supplied from the space heating return and then secondly by a heat exchanger supplied from the DH [5]. This generally results in a lower return temperature to the heat network. Such substations are more complex and so more expensive and will not always be justified but should be considered where DHW demands are high.

Further information and guidance on substation design and controls is given in [5–7].

4.5 Control of bypass flows

An important aspect of the design of the interface is how any bypass flows are set up and controlled. If there is too high a bypass flow then the return temperature in the DH network will rise significantly at times of low demand which will increase heat losses from the return pipe and also result in higher pumping energy than necessary. If there is too low a bypass flow, then it will take more time for heat to be delivered to the building. This is mainly of a concern when instantaneous DHW is used as water would be wasted during the initial draw-off period waiting for the temperature to rise. If there is no bypass flow, then at times of no demand the branch supplying the building will drop in temperature which will lead to lower heat losses from the branch pipe. Such temperature variations on a daily or twice daily pattern may result in additional thermal stress cycling and the long-term life (if steel pipes) could be impacted (see section 5.7).

Although bypasses can be installed within the DH network it is often more convenient to use the building interface as the location for the bypass equipment as access for maintenance will be straightforward. The exception is bypasses at the top of risers in multi-dwelling blocks.

The following are examples of typical approaches to bypass control.

Fixed bypass

A fixed bypass at each building interface will maintain a minimum flow in the network at all times. It will be important to include a DPCV so that the flow does not increase if the pressure difference in the network rises under part load. The fixed bypass will lead to higher return temperatures than necessary and so should be avoided. Even what might appear to be a negligible flow of 5% of peak flow can have a significant effect as for much of the year flow rates will be much lower than the peak (see figure 2.14) [8].

Temperature-controlled bypass

This method uses a control valve linked to a thermostat in the flow pipe so that the bypass is only activated if the flow temperature falls below a set point. This means there will be no bypass unless needed (i.e., at times of no demand from the building) thus ensuring return temperatures are minimised.

Dwelling DHWS

For dwellings using instantaneous hot water production it is necessary to maintain some flow in the local DH circuit otherwise the time to receive hot water at the tap would be too long which would not be acceptable to the customer and would waste water. Two methods are used:

- *Standby flow.* HIUs which include a temperature-controlled bypass to achieve the necessary bypass flow to give acceptable delivery times without contributing significantly to raising return temperatures. This approach, referred to as standby flow or keep warm facility, needs to be carefully designed and is therefore part of the voluntary HIU testing regime developed by the industry [3] (see 4.3.2 above). In many cases the HIU keep warm facility will be sufficient to maintain the network temperature and no additional bypasses will be required.

- *Riser bypass.* In apartment buildings where the design has resulted in very short branches from the riser to the HIU a temperature-controlled bypass may be used at the top of the riser to keep the temperature in the flow riser sufficiently high to meet the demand of the HIUs. This means that any heat losses in the branch to the HIU can be reduced at times when there is no demand for heating and hot water in this branch can be allowed to cool down.

Bypasses on long branches

Some buildings may be connected by long branches off the main network and have either minimal hot water demand or use hot water storage or have well defined operating hours, e.g., office buildings. To reduce heat losses from the branch supplying these buildings a bypass arrangement could be omitted. The long response time would just be accommodated by allowing extra time for heating up after say a night setback. However, this will lead to temperature variations in the buried pre-insulated pipe that will cause stress cycles which, if steel pipe is used, may lead to premature failure from low cycle fatigue (see section 5.7). The same concerns are not present with plastic carrier pipe systems as the compressive stresses from restricted thermal expansion are much lower.

4.6 Interfaces for district cooling

Many of the same principles described above apply to DC systems, including the need to control temperatures, flow rates at cooling devices (AHU's FCU's), and to limit bypass flows.

The main difference with DC is that there is a more significant penalty associated with introducing hydraulic separation as the temperature range is already very restricted. A reduction in DC flow temperature from say 6 °C to 4 °C to allow for a 2 °C drop across a heat exchanger will result in a lower efficiency of the chillers. A direct connection is, therefore, the preferred option. However, similar concerns over water quality and pressure limitations as described for DH may mean that an indirect connection is necessary.

If indirect connection is used the temperature difference across the heat exchanger should be as small as economically possible; the IEA Guidelines [9] recommend 0.8 °C or less.

The heat gains to the DC network will be of less importance than heat losses for DH, however, the need to control pumping energy is more important as this energy adds to the cooling demand, whereas for DH it reduces the heating demand.

Bibliography

[1] CIBSE 2020 *Heat Networks: Code of Practice for UK, CP1.2* (CIBSE)

[2] CIBSE 2021 *Guidance Note: Domestic Hot Water Temperatures from Instantaneous Heat Interface Units* (CIBSE)

[3] Building Engineering Services Association (BESA) 2023 *Technical Standard for UK HIU Test Regime*, 3rd Edition 2023 BESA (https://thebesa.com/heating-interface-units)
[4] Building Engineering Services Association (BESA) *Website for HIU Test Data* (https://thebesa.com/heating-interface-units)
[5] Energiforetagen 2016 *District Heating Substations, Design and Installation, Technical Regulations F:101* (Energiforetagen Sverige) p 9
[6] Parsloe C 2021 *Design Guide: Heat Networks* (CIBSE)
[7] Svend F and Sven W 2013 *District Heating and Cooling* (Studentlitteratur) ch 9
[8] Martin C 2016 Energy efficient district heating in practice—the importance of achieving low return temperatures *CIBSE Technical Symposium* (https://cibse.org/what-s-on/cibse-technical-symposium/past-papers-and-case-studies-archive/2016-papers)
[9] Calderoni M *et al* 2019 *Sustainable District Cooling Guidelines* (IEA DHC/CHP Report) p 91 (https://www.iea-dhc.org/the-research/annexes/annex-xii/annex-xii-project-05)

IOP Publishing

An Introduction to District Heating and Cooling
Low carbon energy for buildings
Paul Woods, MA MSc CEng FIMechE FEI FCIBSE

Chapter 5

Heating and cooling distribution systems

5.1 Network routes and civil engineering

In developing the design of a DH/C system, the likely starting point is a map showing the location of customers and the location of one or more heat production sites.

The primary aim in selecting routes will be to minimise the total length of the network as this will reduce capital costs and heat losses. In addition, there will be a need to plan strategically to enable future expansion of the network and to connect to more remote heat sources. There are likely to be a number of constraints meaning that the optimal route cannot always be selected. These include:

- Places where the density of existing buried utility services is very high
- Road junctions or major roads where traffic disruption would be too great
- Underground structures which obstruct the route, e.g., subways
- Rail and major road crossings where consents will be difficult to obtain

In some instances, a route across parkland or other green space may be preferred as the cost of civil engineering work will be lower even though the route is longer.

It is essential to carry out pre-construction research on existing services and the history of the area to identify potential underground obstructions. This can be followed up by surveys using ground penetrating radar and at critical points trial holes to identify location and depth of services.

These early surveys will contribute to safety during construction. HSG7 [1] defines three elements of a safe system of work:

- Planning the work
- Locating and identifying buried services
- Safe excavation

Buried services can be identified by a number of means and the standard PAS 128:2022 [2] provides the framework for this work. The aim of this standard is to 'lead to more effective planning and safer execution of works as well as less conflict, delay, unnecessary work, damage to third party assets, utility service disruptions and redesigns'.

Extensive guidance on the installation of buried services and how to liaise with other utility providers is given by the industry's trade body Streetworks UK (http://streetworks.org.uk). A recent initiative to map all buried services on a common platform, the National Underground Asset Register, has commenced. For more details see https://www.gov.uk/guidance/national-underground-asset-register-nuar

The main requirements for the civil engineering work associated with DH/C installation are covered by the New Road and Street Works Act [3] and the associated documents that must be followed. It will be necessary for a license to be obtained from the Highways Authority and for discussions to be held in good time about the proposed route. A traffic management plan needs to be prepared and agreed with the Authority. Further information is provided in Objectives 4.1, 4.2 and 4.4 in CIBSE CP1.2 [4].

5.2 Energy analysis of DH/C systems

The heating and cooling networks between the production plant and the customers consist of insulated pipework and a pumping plant which generates the flow. The energy balance of the system (according to the first law of thermodynamics) is given by:

Heat Supplied to Customer = Heat Produced + Electricity for Pump − Thermal Losses

or

$$\begin{aligned}\text{Energy used in transporting heat} &= \text{heat produced} - \text{heat supplied}\\ &= \text{thermal losses} - \text{pumping energy}\end{aligned}$$

In other words, the pumping energy offsets the thermal losses as it is converted to heat within the pipe due to friction. Some of the pumping energy is converted to heat within the pump itself which heats the water. A small amount is lost in the pump's electric motor and inverter which is normally dissipated to air. Some heat is lost from the pump casing to air which can be limited by insulation. Figure 5.1 illustrates the energy balance.

When estimating thermal losses for the purposes of an economic calculation this offsetting is often neglected as it is small in relation to the heat loss for small diameter networks. For larger diameters, however, the thermal loss is smaller relative to the volume flow and the two energy flows can be comparable. In some cases of very large transmission mains the pumping energy can exceed the thermal losses at times of high flow rate, so that the temperature rises along the pipe. This is illustrated in figure 5.2 where the annual pumping energy has been calculated assuming a 20% load factor (i.e., annual pump energy = 20% × peak pump power × 8760) with the peak pipe velocities taken as the recommended figures for feasibility level design in CIBSE CP1.2 [4]. It can be seen that the offsetting becomes significant for diameters above 200 mm.

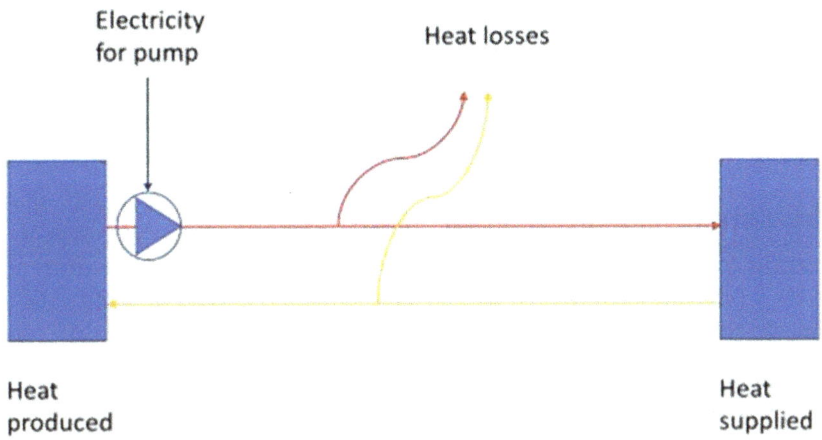

Figure 5.1. Energy balance in a heat network.

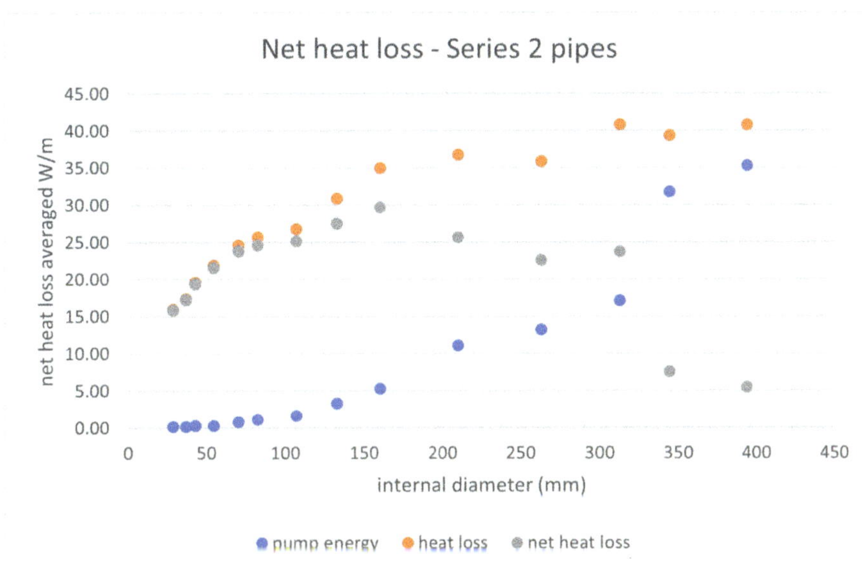

Figure 5.2. Net heat loss for Series 2 pipes taking account of pumping energy.

For DC systems, the same principles apply, however, as it is cooling energy that is being transported the energy flow is in the opposite direction from the customer to the production plant. This means that the pumping energy is additive not offsetting. As the temperature difference between the DC circuit and the ambient is smaller, and the volume flows are greater because of the smaller circuit ΔT, the thermal gains are less in relation to the pumping energy, and it is important to include the pump energy in any calculations of temperature rise between energy centre and the buildings supplied.

The energy balance is:

Cooling energy produced at EC = cooling energy removed from customer + network gains + pumping energy

Energy required for transport of DC = cooling energy produced—cooling energy from customer = network gains + pumping energy

5.3 Pressure drops and efficient pump design

For heating or cooling to be transported between the production plant and customers a flow of water must be established by creating a pressure difference to overcome friction in the pipes and provide the minimum pressure difference at the customer.

Although pumps may be installed within plantrooms located within the network it is usually more practical to install the pumps at the energy centre.

The pressure can be represented in N m^{-2} but it is often more convenient to represent as the head, i.e., equivalent pressure of height of a water column.

$$\text{Pressure} = h\,\rho g \tag{5.1}$$

For density of 975 kg m^{-3} and gravitational acceleration of 9.81 m s^{-2}, 1 m of water is equivalent to a pressure of $1.0 \times 975 \times 9.81 = 9564$ N m^{-2} or 0.095 64 bar.

The pressure distance diagram in figure 5.3 shows how the pressures vary between the energy centre and the furthest customers. This is drawn for the maximum heat demand case—when the flow rates and hence pressure drops will also be at maximum, and the minimum demand case—when the pressure drop in the network is negligible. The differential head shown at A is the minimum required at the customer which is typically maintained by a differential pressure control valve (DPCV). The head required to overcome the flow and return pipe pressure drop and the minimum customer requirement is shown as P.

The diagram represents the pressure changes associated with the flow of water but also the height of the pipe in the network. In a small system or within a low rise

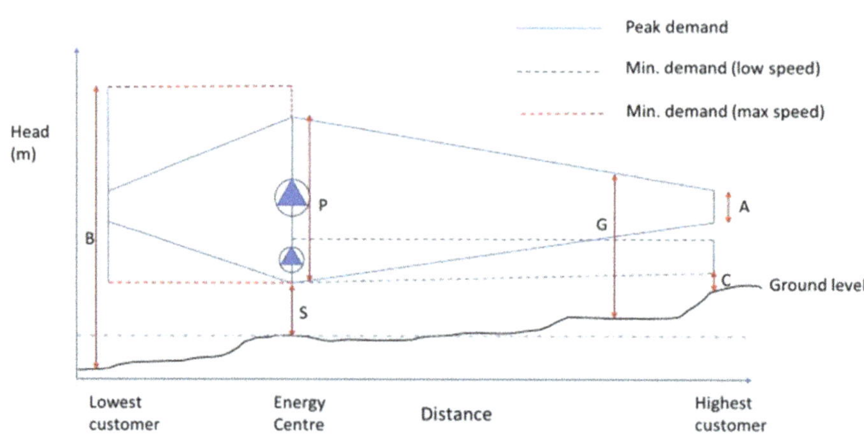

Figure 5.3. Pressure distance diagram.

building this is often not considered, however, in a large network covering significant changes in ground level or where tall buildings are to be connected it is important to take this into account.

On the diagram the ground elevation is also plotted. The pressure measured will reduce as the height of the ground increases and increase where the ground level drops. A pressure gauge installed on the flow pipe at point G will measure a lower pressure as a result of the higher ground level.

The pump pressure to transport the heat to a customer at the top of a hill is no more than that needed for flat ground. This is because the pressure needed to lift the water to the high point in the flow is equal to the pressure created in the return pipe from the height of water above. The pump only needs to overcome the friction in the pipes and customer interface.

Static pressure

The reduction in total pressure as a result of a gain in height is important when setting the static pressure or minimum pressure in the circuit (the position to be maintained when the pumps are off).

To avoid cavitation in valves it is normally a requirement to set the minimum pressure at about 0.5 bar (50 kPa) or 5 m above the pressure at which water would change state to steam at the maximum operating temperature (often taken to be 100 °C). So, in the diagram, the static pressure S must be set to maintain C greater than 5 m gauge.

When connecting very high buildings the static pressure could become very high if the building is directly connected to the network. This could result in higher costs for components in the energy centre and be too high for direct connection of other buildings. A better solution may be hydraulic separation (indirect connection) of the high building (see section 4.2).

Maximum pressure at customers

The maximum pressure that interface equipment, or in the case of direct connection, the customer's heating system will have to accept can also be determined from the diagram. The worst case of maximum pressure at the pumps will be when these are running at maximum speed but with low flow as then the pressure is higher than the pump's design point. If the customer's site is below the level of the energy centre then the maximum pressure is greater than the sum of the static pressure and the pump pressure, as shown as head B on figure 5.3.

If the pressure at customers is too high, then an indirect connection will be needed. Alternatively, a district level heat exchanger substation could be used to hydraulically separate low-lying areas served by the network.

Design of pumps

Although the pumps are often sized to deliver the maximum flow rate to all customers simultaneously, in some cases diversity between customers may mean that the pump sizing can be reduced.

However, the periods when peak demands occur are typically for only a few hours a year and for the majority of the time the demand and hence the required flow rate is much less. This means that the required pressure drop is also much reduced.

Pump power is given by:

$$P = Q \times \Delta p/\text{efficiency} \tag{5.2}$$

where P is the input power required (W), Q is the volume flow rate ($m^3\ s^{-1}$) and Δp is the pressure rise ($N\ m^{-2}$). The pump efficiency is quoted by manufacturers and should include the efficiency of the pump itself, the electric motor and the inverter drive if used. The pump should be selected to operate in the part of the pump curve where efficiency is highest.

The performance of a pump is defined by the pump curve. These are provided by the manufacturer usually showing the relationship between head (pressure rise as m of water) and flow rate ($m^3\ h^{-1}$). The flow rate is determined by the resistance in the circuit the pump is supplying. For a given flow rate the pump will deliver the head shown by the curve.

The required pressure for the network (flow and return) at any given flow rate can also be calculated and will follow an approximately parabolic curve as pressure from pipe friction is approximately proportional to velocity squared. A minimum pressure difference is required at the end customer so this can be added to the network pressure characteristic to give a DH system curve.

Combining the pump curve and the system pressure drop curve together will indicate where the pump will operate at the peak demand condition.

To illustrate this, figure 5.4 shows a pump curve at 100% speed and the DH system curve which intersect at point A which is the peak demand condition. The

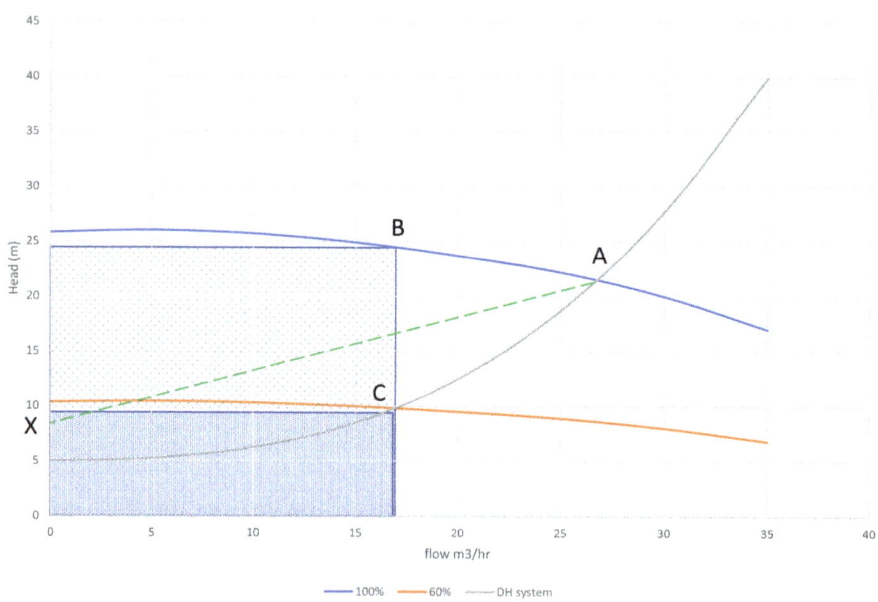

Figure 5.4. Pump curve diagram.

system curve assumes that a 5 m head difference is required at the customer connection for all levels of demand hence the system curve crosses the y-axis at 5 m.

As demand reduces the flow rate is reduced as control valves operate at the customer to restrict the flow of water (assuming a two-port control valve system, see section 4.2.4) and as the flow rate reduces the operating point moves up the curve from A to B. Now the pressure drop is higher than that required to overcome the friction in the pipe. The excess pressure is lost in the control valves which are closing down.

The pumping energy is almost all turned into heat energy through friction but as electrical energy is typically 3–10 times as costly as heat energy it is important to minimise it.

The pumping power for the operation at B can be calculated from the formula (equation (5.2)). This shows that the power is proportional to pressure x flow rate which is represented on the diagram as the lightly shaded rectangular area (with the top-right corner at B and bottom-left corner at the origin). The pump efficiency will be less than at the design point as this would have been selected for maximum efficiency.

To improve the efficiency at part-load it is necessary to either use a smaller pump or to operate the main pump at reduced speed.

The 60% curve shown on figure 5.3 represents the main pump running at 60% speed but it could also represent a second smaller pump. The operating point is at C where the system curve intersects the new pump curve.

The pumping power for the operation at C is proportional again to the area of the rectangle with darker shading (with the top-right corner at C). It can be seen that the power required will be less than half the power for the fixed speed case.

For a variable speed pump there will be additional losses in the inverter drives and so the overall pump efficiency will be lower at 60% speed. To cover the full range of demands it is common to have both variable speed drives and multiple pumps of different sizes and heads. In the summer period especially, a smaller pump would be justified as both volume and required head will be much lower than in the heating season.

Pump controls will aim to reduce the head from the pump to match the required head as closely as possible. Ideally differential pressure signals will be obtained from within the network at multiple points so that the speed can be adjusted to maintain set differential pressures. Alternatively, the pressure difference at the pump can be set to follow a pre-determined relationship with flow rate which approximates to the system curve shown as the green line between A and X on figure 5.3 such that the pump will always deliver a head more than the required system head. This is known as proportional control.

Standby pumps

Pumps are frequently specified as 100% run and standby. This can be expensive and does not deal with the part-load requirements well so an alternative of 2 at 50% could be considered. When one pump fails this does not mean that only 50% of the flow will be delivered as the pump will be operating at a different point on its curve. Typically, one of the 50% pumps can deliver around 70% of the peak flow rate which may well be enough under most operating conditions. The drawback of this arrangement is that if the failure occurs at a time when a higher flow than 70% is needed the heat supply would not be evenly distributed to all customers.

Pump bypasses

If a pump operates against a closed head the water within the pump will heat up and could turn to steam which would damage the pump. So a minimum flow rate needs to be provided at all times. Typically, this is achieved from bypass flows within the network (see section 4.5) but if necessary a bypass around the pump or between the pump and the return can be provided to be opened only when the flow rate is very low.

5.4 Heat losses

5.4.1 Quantifying heat losses

Some of the energy transported to the customer is lost through heat losses to the environment.

Heat is transferred through the insulation and the ground to the ambient air. Pre-insulated steel pipes supplied to EN253 [5] need to meet minimum standards for heat loss. The insulating effect is achieved by having pockets of gas trapped in small cells within the polyurethane foam structure. Typically, pre-insulated pipes are manufactured with three different thicknesses of insulation. Manufacturers publish heat loss rates for each level of insulation which take account of the insulating effect of the ground to the ambient air (see figure 5.5). The lowest heat loss rate is obtained when using twin pipes (see section 5.6.3).

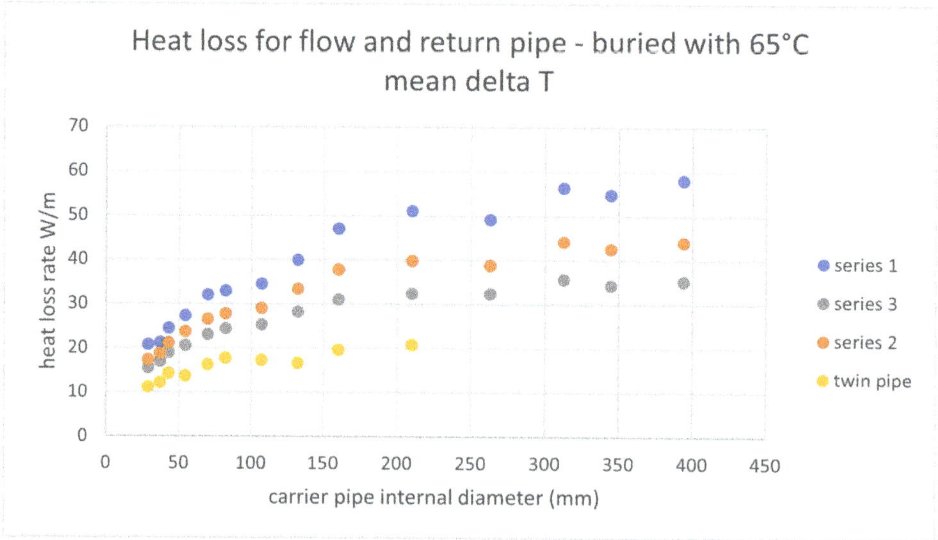

Figure 5.5. Heat loss for flow and return pipes for three standard insulation thicknesses and medium thickness twin pipes (assumes $T_{\text{mean}} - T_{\text{ambient}} = 65$ °C), typical data from manufacturer's catalogue (Powerpipe), assumes thermal conductivity of ground at 1.5 W mK^{-1}.

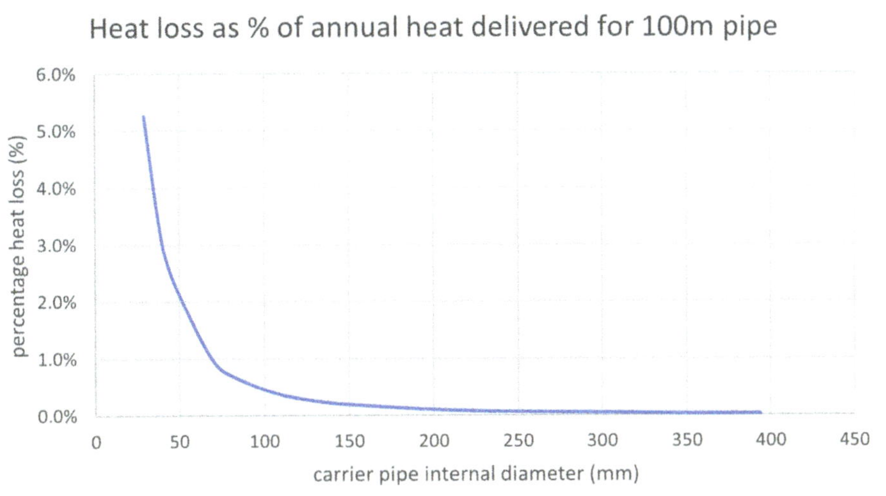

Figure 5.6. Heat loss as a percentage of heat delivered for a range of pipe sizes (100 m length and Series 2 insulation).

The annual heat loss can be related to the annual heat that would typically be delivered by each pipe size assuming peak velocities as recommended in CIBSE CP1.2 [4] and a load factor of 40%. These are shown in figure 5.6 expressed as a percentage for a pipe length of 100 m. Heat gains from friction are ignored. It can be seen that the percentage heat loss is much higher for smaller diameters but reduces to very low levels for larger diameters. This is because the surface area increases in proportion to the diameter whereas the energy flow rate increases in relation to both the square of the diameter and the velocity of flow; higher design velocities are used with larger pipes.

The heat loss will depend on the temperature difference between the fluid in the pipe and the ambient temperature and as both of these will vary through the year an hour-by-hour calculation would be more accurate. However, for most purposes taking annual averages of ambient temperature and flow and return temperatures will enable the annual heat loss to be determined with sufficient accuracy.

For DH systems operating with a wide ΔT, the heat losses from the return pipe will be less than for the flow pipe as the temperature in the return is lower. Therefore, there would be a case for using a higher level of insulation in the flow pipe than the return pipe although this would make managing materials on site more complex.

There are normally three key calculations to be carried out in relation to heat losses:

- The annual heat loss as this amount of heat will need to be produced at the energy centre and allowed for in the operating cost model.
- The temperature drop to the furthest consumer at times of low demand as this will determine the minimum supply temperature from the energy centre.
- The temperature drop to the furthest consumer at times of peak demand as this will determine the supply temperature at times of peak demand at the energy centre.

For DC systems the thermal gains are lower as the temperature difference with the ambient is less and again most of the gain will be in the flow pipe.

For ambient networks the temperature difference will be small but the pipes will still need to be insulated to at least Series 1 level especially when running above ground exposed to ambient air temperatures.

For a given network design the thermal losses will be determined by the flow and return temperatures and the lower these are the lower the heat losses (for a given pipe size). Hence the flow temperature is often reduced at times of part-load when the customer's needs can be satisfied at lower temperatures and keeping the return temperature as low as possible is a key objective for the design of the building interfaces and the control of bypass flows (see sections 3.2.3, 3.3.3 and 4.5).

The major differences in heat losses between different projects and designs will be a function of:

- Network length
- Insulation thickness
- Operating temperatures
- Pipe sizing

The latter two are discussed in section 5.5 and chapter 7.

5.4.2 Special case of multi-residential blocks

Heat losses from distribution pipework within a block of flats are of a particular concern because:

- The heat losses lead to unwanted internal gains to the building in summer contributing to over heating.
- Poor design of the layout of pipework can lead to unnecessarily long lengths of pipe leading to high losses.
- Pipe sizes are generally small so the heat loss is significant in relation to the annual heat supplied.
- In many cases the annual heat supplied is low due to partial occupancy and low load factors.

The heat losses in winter contribute to heating the block but are before the heat meter in each flat. As a result, the heat taken by each flat is less than it would have been but to compensate for this the unit price needs to be increased. This can show DH in a poor light compared to other heating systems.

The target for heat losses in CIBSE CP1.2 [4] is set at 100 W per dwelling. So this effectively limits the length of pipework to around 10 m per dwelling even with high levels of insulation. Figure 5.7 below shows how different riser and lateral

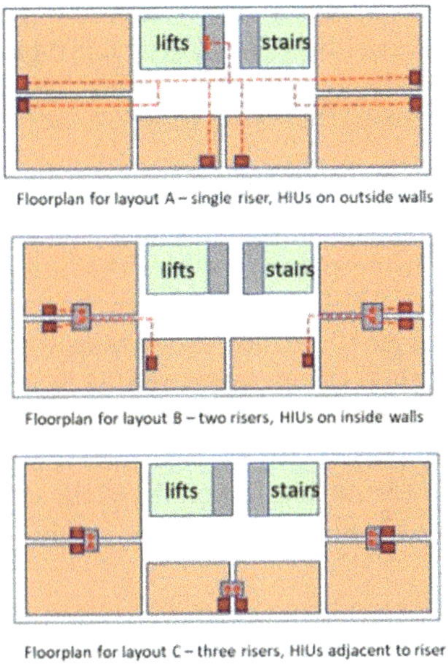

Layout	A	B	C
Riser length (m)	3	6	9
Laterals length (m)	81	30	11
Total length (m)	84	36	20
Total length per dwelling (m)	14	6	3.3

Figure 5.7. Possible pipe layouts in a typical floor plan for a block of flats.

distribution layouts and positions of HIUs within a block can result in very different lengths of pipe per dwelling. Early discussions with the architect on the location of risers and HIU's are needed to achieve an optimal design to minimise heat losses.

5.4.3 Impact on economic analysis

When assessing the impact of heat losses on the economy of the scheme it is important that the heat loss is valued at the cost of heat produced not the heat selling price. The only effect of higher heat losses is an increase in the heat production cost.

For example, if the cost of heat production is around 5 p kWh^{-1} and the losses per dwelling are 100 W or 876 kWh p.a. then the cost of the heat loss is £44 p.a. per dwelling and not the £88 p.a. per dwelling that would be erroneously calculated using a heat selling price of 10 p kWh^{-1}.

It is worthwhile evaluating an optimal insulation thickness, taking account of the capital cost for the pipes, the cost for installation, which depends on trench width, and the cost of heat production over the life of the scheme. The optimal insulation thickness will vary with pipe size and between flow and return.

5.4.4 Environmental impact

The environmental impact of heat losses needs to consider how the production plant would operate if the heat loss was higher or lower and then calculate the impact on CO_2 emissions. It will be more important to try and reduce losses at times of peak demand when the marginal plant supplying the system may be gas-fired boilers than in summer when the demand is met by say gas-fired CHP plant or heat pumps.

5.4.5 Intermittent operation to reduce heat losses in branches

Where there are long branches to relatively small customers the heat losses from these branches can be significant, e.g., for a housing network where the houses are set back from the road containing the street main. Section 4.5 discusses the potential benefits of intermittent operation by the customer to reduce heat losses from the branch pipe (by limiting bypass flows). This method of operation results in temperature cycling of the branch pipe which with steel systems increases the risk of failure from low cycle fatigue (see section 5.7.4). With plastic carrier pipes this issue does not arise as the expansion stresses are much lower in relation to the yield strength and operating at lower temperatures for significant hours in the year would in any case extend the life of the pipe.

5.5 Pipe sizing and optimisation

5.5.1 Cost elements

For a given volumetric flow and operating temperatures there will be an optimum pipe size to transport the energy that seeks to minimise the sum of:

- The capital cost of the pipe—which increases with size
- The cost of heat losses from the pipe—which increases with size
- The cost of pumping energy to overcome friction—which decreases with size

5.5.2 Capital costs

Capital costs for the pipeline itself are related mainly to the diameter of the pipe as this determines the amount of material used. The civil costs are related to the volume of the trench for excavation and backfill which is related to the diameter squared as the width and depth of the trench increases with diameter. It is also found that there are high fixed costs associated with the construction work.

So a suitable equation is of the form:

$$C = a + b\, d^n$$

where C is the capital cost per meter of trench length, a and b are constants, d is the internal pipe diameter (in mm) and n is a constant.

Costs will need to be estimated based on recent tender returns for the locality concerned. For the purposes of developing figure 5.8 and table 5.1 the following formula has been derived for steel carrier pipes to EN253. Other cost curves will apply to plastic carrier pipes

$$C(\pounds/m) = 550 + 0.4d^{1.4}$$

The capital cost can be annualised according to discounted cashflow techniques where for a 25 year life and a discount rate of 3.5% the annualisation factor is 16.5, i.e.,

$$\text{Annualised capital cost} = \text{capital cost}/16.5$$

5.5.3 Pumping energy cost

For a given heat demand Q (kW) the mass flow rate required can be calculated for the selected temperature difference (ΔT) between flow and return.

$$Q = \dot{m}\, c\Delta T \tag{5.3}$$

where $\dot{m}$ is mass flow rate (kg s^{-1}), c is specific heat (kJ kg^{-1} K), and so

$$\dot{m} = c\Delta T / Q$$

The equation relating mass flow rate to velocity and cross-sectional area is the continuity equation which represents conservation of mass:

$$\dot{m} = \rho\, A\, v \tag{5.4}$$

where: $\dot{m}$ is the mass flow in kg s^{-1}, ρ is the density (kg m^{-3}), A is the cross-sectional area of the pipe (m^2) and v is the flow velocity (m s^{-1}) hence for a given diameter, A is calculated and hence v can be calculated from

$$v = \dot{m}/\rho\, A$$

To calculate the pumping energy, the pressure drop needs to be determined from the Darcy–Weisbach equation, see [6]:

$$\Delta p = \lambda l/d.\ (\rho v^2/2) \tag{5.5}$$

where λ is the friction factor, l the length of pipe, d the internal diameter.

From this equation it can be seen that the pressure drop is proportional to the length of the pipe and to the square of the velocity. As the velocity is inversely proportional to the square of the diameter (equation (5.4)) the pressure drop is inversely proportional to diameter to the fifth power.

In 1944 Moody presented data on pressure drops in graphical form (the Moody chart) which shows the friction factor in relation to Reynolds Number (Re) and the relative pipe roughness (k d^{-1}).

The Reynolds number is a dimensionless number which indicates the level of turbulence in the flow:

$$\text{Re} = \rho v d/\mu$$

where μ is dynamic viscosity (N m^{-2} s^{-1}).

A typical roughness figure k that may be used for steel DH pipes is 0.1 mm. Table 4.1 in CIBSE Guide C [7] provides 0.04–0.1 mm for new seamless steel pipe.

Roughness for plastic pipes is typically assumed at 0.003 mm and is likely to remain at this figure as significant fouling is less likely to occur.

The Moody chart was based on the formula proposed by Colebrooke and White which is the most accurate representation of λ but requires iterative methods to solve it.

Various simpler equations have been developed that represent the data on the Moody chart.

The equation proposed by Moody in 1947 was:

$$\lambda = 0.0055\{1 + (2.10^4 dk + \text{Re}.10^6)\}^{0.33} \text{ valid for } k/d < 0.01 \text{ and } 4000 < \text{Re} < 5.10^8$$

An equation proposed by S E Haaland in 1983 [8] is:

$$1/\lambda^{0.5} = -1.8 \log_{10}\{6.9/\text{Re} + (k/3.71\ d)^{1.11}\}$$

This allows λ to be derived directly within 1.5% error compared to the Moody chart. This formula is the one recommended in CIBSE Guide C [7].

In addition to the pressure drop for the pipe, additional losses will occur at bends and branches. For DH systems these are usually relatively small but still require calculation in accordance with standard sources, e.g., CIBSE Guide C [7].

Once the pressure drop has been calculated (for both flow and return) the pumping power can be calculated from equation (5.2) however this is only for the design flow and a load factor needs to be applied which takes account of the demand pattern over the year which results in variations in both volume flow and required pump pressure. The load factor depends on how closely the pump speed is varied with demand to minimise excess pressure and the losses in the pump and inverter. Also, the volume flow rate will change not just with demand but also with ΔT which may vary if the flow temperature is reduced at part-load. Therefore, an hour-by-hour simulation is needed to give an accurate calculation of the pumping energy.

From a typical simulation and assuming the use of variable speed pumps and constant flow temperature a load factor of 20% has been derived so:

$$\text{Annual pump energy (kWh)} = \text{pump energy at peak (kW)} \times 8760 \times 20\%$$

5.5.4 Heat losses

Heat losses can be calculated for each diameter for the selected operating temperatures. Where these are varied over the year an hourly average temperature will be needed. For this illustration a flow temperature of 85 °C and a return temperature of 55 °C with an ambient temperature of 10 °C were assumed (a mean ΔT of 60 °C). In the future, it will be more common for lower temperatures to be used under part-load conditions where heat pumps are used.

As seen in section 5.2, the friction in the pipe results in heat gain to the water in the pipe which should be taken into account.

5.5.5 Total annualised costs

To calculate the annual operating costs from the above, energy prices need to be assumed for the electricity for the pumps and the heat production. As these vary over the year an hour-by-hour calculation may be necessary for accuracy but for this illustration the following were assumed:

Electricity cost: 15 p kWh^{-1}

Heat production cost 5 p kWh^{-1} (e.g., from a heat pump with a CoP of 3)

Figure 5.8 shows how the cost elements described above are added to give a total annual cost with the minimum cost defining the optimal diameter. It also shows that undersizing of pipes can lead to a rapid increase in cost (as the pumping energy is inversely proportional to the diameter to the fifth power) but oversizing is less critical. This is helpful when sizing for future expansion as although the capex will be higher this will initially be offset by lower pumping energy costs.

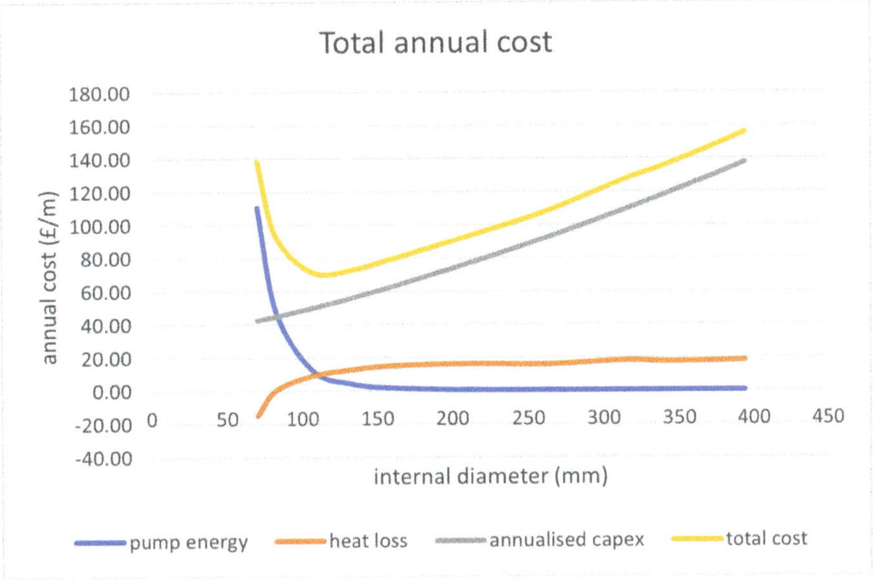

Figure 5.8. The principle of optimal pipe sizing to minimise total annual cost.

5.5.6 Optimal sizing on a per m basis

For a range of flow rates the optimal diameter and hence optimal velocity has been calculated for each using the same assumptions as given in the preceding sections. As pipe sizes are only available in discrete sizes these velocities should be seen as a guide to optimisation and for use in initial sizing. The heat losses were offset by the pumping energy. Energy prices were assumed to be constant as in section 5.5.5. The results are presented in table 5.1 and compared with the recommended velocities to be used at feasibility stage taken from CIBSE CP1.2 [4]. This shows that the optimal velocity is similar to CP1.2 for larger diameters but higher for smaller diameters.

The pressure drops per m at the smaller diameters are significantly greater and it is often the total pressure constraints on the network that will govern the design and lead to lower velocities being selected for the smaller pipes. Similar results were obtained by Oliver Martin-Du Pan [9].

Table 5.1. Comparison of minimum cost velocity and recommended sizing velocities in CIBSE CP1.2 [4] for steel pipes.

Pipe NB (mm)	Velocity in CP1.2 (m s^{-1})	Pressure drop per m for CP1.2 velocities (Pa m^{-1})	Optimum velocity calculated by model (m s^{-1})	Pressure drop per m for optimum velocities (Pa m^{-1})
25	0.9	392	1.3	869
32	0.9	286	1.4	707
40	1.0	291	1.4	563
50	1.0	216	1.4	419
65	1.3	262	1.5	336
80	1.4	248	1.5	295
100	1.5	205	1.8	299
125	1.8	226	2.0	283
150	2.0	220	2.2	272
200	2.4	226	2.6	264
250	2.4	172	2.7	219
300	2.5	151	2.8	185

5.5.7 Software

To assist the designer, specialist software is available to carry out calculations of pressure drop and temperature drop across a network. Some software relies on the use of defined velocities for pipe sizing as in table 5.1 other software carries out an economic optimisation using the principles outlined above. The following are some of the commercially available products:

Fluidit Heat—from Fluidit Oy, Finland [10]
Termis—from Schneider Electric [11]
Netsim—from Vitec Energy AB, Sweden [12]

5.6 Technologies

5.6.1 Pre-insulated steel pipes

Until the 1970s, DH and DC systems used conventional pipework and insulation installed in concrete ducts which were designed to be free draining. Provided the ducts were kept clear of water this was a robust system but expensive to install. In the 1970s the concept of pre-insulated pipes became established which could be buried directly in the ground without the need for ducts. The various products became more standardised through the work of the CEN with the first product standard EN253 approved by CEN in 1988. This remains the core standard for pre-insulated pipe systems with the latest revision dated 2019 [5].

The standard pre-insulated pipe consists of a steel carrier pipe surrounded by rigid polyurethane foam insulation and an outer casing of high density polyethylene (HDPE). The insulation is bonded to the steel and the outer casing during the foaming process. This combines the high strength at high temperatures of steel, an efficient layer of insulation and robust protection by the outer casing.

The sections of steel pipe are welded together so that the risk of leaks is minimised and maintenance is not required. An HDPE sleeve is installed over the weld and the cavity insulated.

Joints

There is a risk of external corrosion if ground water penetrates the joint area and so considerable effort has been made to provide insulated joints that will prevent water entering the joint area. Both mechanical couplings and glues have been used to affect a seal but more commonly now the sleeve and the HDPE casing are joined using electrofusion.

The joint area is insulated by injecting foam into the void. This ensures there are no gaps in the insulation and any water ingress will take longer to reach the steel pipe. There is a need for holes to be made in the joint casing to allow the foam liquid to be injected which then have to be carefully sealed.

Similar systems are used for DC where the main risk is condensation occurring on the steel pipe and the sealed system is equally effective in preventing this (see section 5.11). The pipe insulation may be thinner than for DH pipes.

The EN standard covering requirements for joints is EN489 [13]. This includes a test known as the sand box test where the joint is placed in a box of sand to simulate the ground conditions and the pipe is subject to movements to prove that the joint will maintain integrity when it moves in the ground as a result of thermal cycling.

Bends and branches

Manufacturers produce pre-insulated bends and branches which simply need to be joined to straight lengths using the same approach as connecting two straight lengths together, i.e., welding of carrier pipe, fitting a casing sleeve and insulating the cavity. The relevant EN standard is EN448 [14].

Hot tapping of main lines whilst they remain in operation is also available. Normally the diameter of the branch is less than half that of the main line but hot tapping of a branch with a diameter of up to 95% of the diameter of the main line is possible.

Surveillance system

Because of the potential for external corrosion damage if ground water ingresses into the joint area, a surveillance system (also known as a leak detection system) is incorporated in the pipe. Typically, this consists of two copper wires (one tinned) embedded in the insulation. If the insulation becomes damp either from ground water or DH water then the resistance between the wire and the steel carrier pipe will change and this can be identified from measurements at the end of the pipeline and an automatic alarm raised. The location of the fault can be determined by sending a high frequency AC current down the wires and measuring the impedance and from this deduce the position of the fault to an accuracy of about 0.5% of the system length. An early warning of a fault which can then be rectified before corrosion damage occurs can save significant costs and avoid the need for shutdowns for repairs.

It is recommended that early design team discussions are held to determine the capability and suitability of the proposed system and the location of required measuring points. After installation the system needs to be calibrated to enable accurate fault location to be achieved.

The relevant EN standard specifying minimum performance levels for surveillance systems is EN 14 419 [15].

5.6.2 Plastic carrier pipes

As an alternative to the steel carrier pipe systems, plastic has also been used for the carrier pipe for several decades. The main disadvantage of most plastics is that their composition will change when under stress resulting in creep failure; the rate of this change is both temperature and pressure dependent. At high temperatures the lifetime of plastic materials is limited. Three types of plastic that are suitable for operating at elevated temperatures are:

- Polybutylene which can be fusion welded (PB).
- Cross-linked polyethylene which is joined using mechanical couplings (PE-X).
- Polypropylene which can be fusion welded (PP-R and PP-RCT).

As DH systems started using lower temperatures other benefits of plastic carrier pipes became recognised:

- The risk of corrosion from either the ground water or DH/C system water is avoided, water treatment is less critical (but may be for other components in the system).
- The pipes are more flexible making them easier to install.
- The pipes can be produced in long lengths reducing the need for joints which reduces cost and failure risk.
- The pipe is less prone to internal fouling and pressure drops are normally lower than the equivalent steel pipe.

The strength characteristics of plastic are such that larger diameters require much greater wall thickness and the benefit of flexibility is then lost. So, in the past most applications have been in smaller diameters especially for the final branches to individual houses. They are particularly effective for long branches across undulating landscaped areas.

Table 5.2. Temperature regime specified in EN15362.

Temperatures (°C)	Cumulative period in 30 years
80	29 years
90	7760 h
95	1000 h
100	100 h

The 100 °C operation is assumed to be under fault conditions.

Table 5.3. Temperature regime from Annex A of EN15362.

Temperatures (°C)	Cumulative period over a year (h)
65	535
70	3720
75	840
80	3500
85	165

Plastic pipe systems are more expensive than steel pipes and the benefits of lower installation costs need to outweigh the extra material cost.

The main disadvantage is that the pressures and temperatures need to be limited to maintain a long life for the system. Pressure is normally limited to 6 bar g although higher pressure variants are available. The standard EN15362 [16–18] requires that the pipe will have a 30 year life under the operating temperature regime in table 5.2:

Annex A of EN15362 [18] contains an alternative temperature regime which by applying Miner's Rule from EN ISO 13 760 [19] can be shown to have a life of over 50 years (see table 5.3). This profile is of interest as it may be more representative of a heat network designed with a maximum temperature of 85 °C and variable operating temperatures as discussed in chapter 7. Lower temperatures are likely to be used in the future as heat pumps become more widely used and so the issue of limited life of plastic systems will become less of a concern.

Diffusion barrier

Oxygen diffusion can occur through plastic leading to an increase in oxygen within the DH/C system water which would increase the risk of corrosion in any steel components in the circuit. The use of a diffusion barrier such as an aluminium foil layer is, therefore, often used to prevent this diffusion.

Insulation of plastic pipes

Because one of the benefits of plastic pipes is the flexibility which also allows long lengths to be produced and delivered as a roll it is also necessary to have flexible insulation. This may be in the form of layers of polyethylene foam insulation. Typically, the heat losses from this type of insulation are higher than for the steel systems and this needs to be considered when comparing the available solutions. The alternative of using polyurethane foam where the insulation is bonded to the carrier pipe will still provide sufficient flexibility for most installations and will have lower heat losses for the same insulation thickness. The EN15632 standard has separate parts for bonded and non-bonded systems [17, 18].

Other flexible pipes
Flexible pipes are also available with metal carrier pipes such as copper and corrugated stainless steel. These pipe systems are covered by Part 4 of EN15632 [20].

Straight pipes
The trend towards lower operating temperatures and the need to reduce the risk of corrosion failure have led to straight rigid plastic carrier pipe systems becoming more widely used and at larger diameters. PB and PP-R and PP-RCT are all offered and these can be welded easily using technologies and skills found in other industries.

At present there is no EN standard that covers the larger diameters of pre-insulated plastic pipe sold in straight lengths as EN15632 is written to cover the flexible pipe systems available in the market.

Reinforced plastic pipes
To overcome the limitations of temperatures and pressures a more recent innovation is the reinforced plastic carrier pipe which uses embedded Kevlar fibres within the plastic. The additional strength means pressures of up to 10 bar g can be accepted.

Embodied energy
The embodied energy associated with plastic production is less than that associated with steel.

5.6.3 Twin pipes

A further option available to designers is the use of twin pipes, i.e., two carrier pipes in one casing. Designs are available for both plastic and steel carrier pipes up to 250 mm NB. The benefits are:

- Lower heat losses (see section 5.4.1)
- A narrower and shallower trench can be used reducing civils costs

The disadvantage is that the installation at bends needs to be in the plane of the pipes whereas with individual pipes the installation level can be varied more easily. Thus, twin pipes tend to be used in new developments where co-ordination with other services can be achieved in advance, rather than in existing roads where the ability to change direction and even use different routes for separate flow and return pipes is advantageous to circumvent existing services. Twin steel pipes are also more difficult to weld due to the limited space between the pipes.

The EN standard for twin pipes is EN15698, Part 1 for the pipe assembly [21] and Part 2 for the fittings [22].

Although most twin pipes are manufactured within a circular casing other shapes are possible to reduce trench width and the amount of materials used.

5.7 Design of steel pipe systems and application of EN13941

5.7.1 The need for stress analysis

In conventional building services, stress analysis of steel pipe systems is unnecessary as the internal pressures are relatively low. However, it is common to consider thermal expansion and to make allowance for thermal movements.

In DH the pressures are also relatively low so thin-walled pipe can be normally used. However, because the friction with the ground constrains the pipe and free expansion is not permitted the thermal stresses induced by temperature change are significant.

In DC both pressures and temperature changes are small and a full stress analysis is not normally required.

For plastic carrier pipes the induced thermal stresses are much lower than the yield stress so a similar stress analysis to steel pipes is not required

EN13941 is in two parts: Part 1 covering design and Part 2 installation (including testing and NDT) [23, 24]. At first reading the standard appears complex however this is because it is designed to cover all applications including systems with high temperatures and large diameters which are unlikely to be commonly used in UK projects. The starting point is the classification of the project into Project Classes according to figure 5.9.

Project Class A involves diameters less than 300 mm and a temperature rise between installation and operation of less than 85 °C and so covers most projects

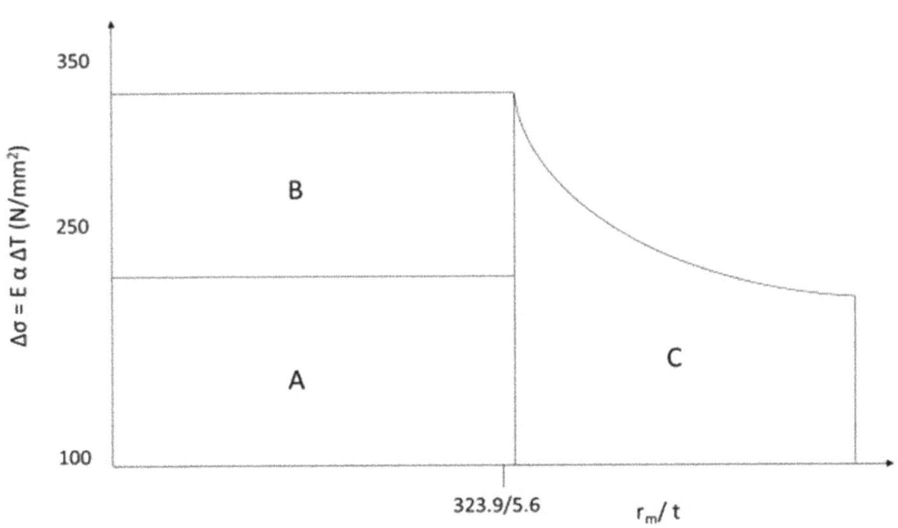

Figure 5.9. Project Classes from EN13941 Part 1 (redrawn).

Table 5.4. Classification of limit states in EN13941.

	Limit states
Force-controlled:	
A1	Internal pressure and external loading from traffic and selfweight of soil, selfweight of pipe and contents
A2	Stress ratcheting
Displacement-controlled:	
B1	Low cycle fatigue
B2	High cycle fatigue
C1	Instability—local buckling
C2	Instability—gross buckling
Other:	
D	Serviceability limits

likely to be built in the UK. Project Class B is for diameters less than 300 mm but with higher temperatures. Project Class C is for diameters larger than 300 mm with any temperature range. In addition to the diameter and temperatures the risk of pipe failure is taken into account when assigning classes. So a small diameter low temperature pipe could still be classed as Class C where there is a high risk of personal or property damage if there is a failure. This might apply for a pipeline crossing a railway for example.

For Project Classes A and B the simplified rules published by manufacturers can be used, provided these have been derived using EN13941. The Project Class also determines the requirements for testing and NDT and the safety factors to be applied in fatigue analysis.

EN13941 Part 1 requires the designer to evaluate the design against a number of failure modes or limit states. These are classified either as force-controlled actions or displacement-controlled actions as in table 5.4.

The principles of the most relevant limit states are outlined below. Reference must be made to EN13941 for any particular design application.

5.7.2 Limit state A1 and A2—stresses arising from internal pressure

Circumferential (hoop stresses)

The circumferential stress in a pipe wall arising from internal pressure is derived by resolving the pressure forces acting in one direction with the tensile forces in the pipe walls, see figure 5.10.

From figure 5.10,

$$\sigma_c\, 2t = p\, d_i$$

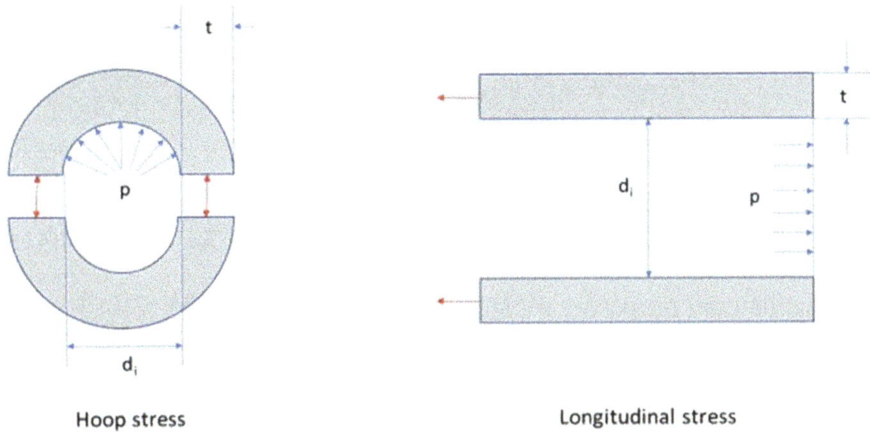

Figure 5.10. Stresses arising from internal pressure.

$$\sigma_c = p\ d_i/2t$$

Longitudinal stress from pressure
This tensile stress arises from the pressure exerted at the end of the pipe, e.g., at a 90° bend. From figure 5.9,

$$\sigma_1\ t\ \pi\ d = p\ \pi d^2/4$$

$$\sigma_l = p\ d/4t$$

Taking typical values of 10 bar g and using a 100 mm NB pipe these stresses are:

$$\sigma_c = p\ d/2t = 1\ 000\ 000 \times 0.1071/2 \times 0.0036 = 14.9.\ 10^6\ \mathrm{Nm^{-2}}\ \text{or}\ 14.9\ \mathrm{Nmm^{-2}}$$

$$\sigma_l = p\ d/4t = 1\ 000\ 000 \times 0.1071/4 \times\ 0.0036 = 7.44.\ 10^6\ \mathrm{Nm^{-2}}\ \text{or}\ 7.44\ \mathrm{Nmm^{-2}}$$

These pressures can be compared to the yield stress of the steel used of 235 N mm^{-2} and so it can be seen that the stresses due to internal pressure are relatively low. The recommended safety factor for design is 1.2 and 1.25 for welds giving total safety factor of 1.5. In principle even thinner walled pipe could be used but this would be impractical to weld and more susceptible to failure from corrosion.

Limit state A2—stress ratcheting
EN13941 includes a limit state associated with stress ratcheting or repeated actions causing high stresses. It states that this limit state can only occur for very high pressures (>20 bar) and large diameters and so is not discussed further here.

5.7.3 Displacement-controlled actions and limit states

Displacement-controlled actions are a result of the stresses introduced from the temperature changes when the pipe is put into operation.

The compressive stress in a fully constrained pipe subject to a temperature rise of ΔT is given by:

$$\begin{aligned} &\text{Stress} = \text{E} \times \text{strain} \\ &\text{Strain} = \Delta L/L \text{ where } \Delta L = L\,\alpha\,\Delta T \qquad (5.6) \\ &\text{So } \sigma = E\,\alpha\,\Delta T \end{aligned}$$

It can be seen that the pipe wall thickness is not included in the equation so the stress cannot be reduced by increasing the wall thickness unlike for the stresses arising from internal pressure.

Using the material properties for steel given in EN13941 at 90 °C of E = 209 kN mm^{-2} and $\alpha = 12.1\ 10^{-6}$ mm mm^{-1} and a $\Delta T = 90$ °C (e.g., installed at 0 °C ambient and an operating temperature of 90 °C) results in a stress of:

$$E\,\alpha\,\Delta T = 209,000 \times 12.110^{-6} \times 90 = 227\ \text{Nmm}^{-2}$$

It can be seen that this stress is close to the typical yield stress of 235 N mm^{-2}.

If the pipe was anchored at each end, the force the pipe would exert on the anchor is given by:

$$P = \sigma\,A$$

So for a typical 100 mm pipe with a cross-section area of 1252 mm^2 the force would be 285 kN or 29 tonnes f. Awareness of the magnitude of this potential force in the pipe is important when considering the design of entries into buildings where allowing movement is critical so that these high forces from constrained expansion do not arise.

The thermal expansion stresses are considered secondary stresses and cannot lead to a collapse failure mode unlike pressure stresses. Nevertheless, designers have sought to reduce the stress levels in the pipes to about 2/3 of the yield stress. Clearly using lower operating temperatures will reduce the temperature rise that is applied. However, for temperature rises greater than about 60 °C, there are three methods available:

(A) Using changes of direction at intervals so that the stresses are limited and some movement of the pipe is permitted

This is the method that has been traditionally used. Where there is a 90° bend it can be assumed that movement in the ground at the bend is relatively unconstrained see figure 5.11. The friction forces from the ground then build up away from the bend so that at point B the pipe becomes locked and the maximum compressive stress given in equation (5.6) above is seen.

To limit the stress to the level at point C, there is a maximum length of pipe that can be installed. Design guides published by manufacturers will define this maximum length of pipe for various configurations and diameters so that the stress is limited to about 60% of the yield stress. Foam cushions can be buried with the pipe to provide material at the bend that does not constrain the pipe movement.

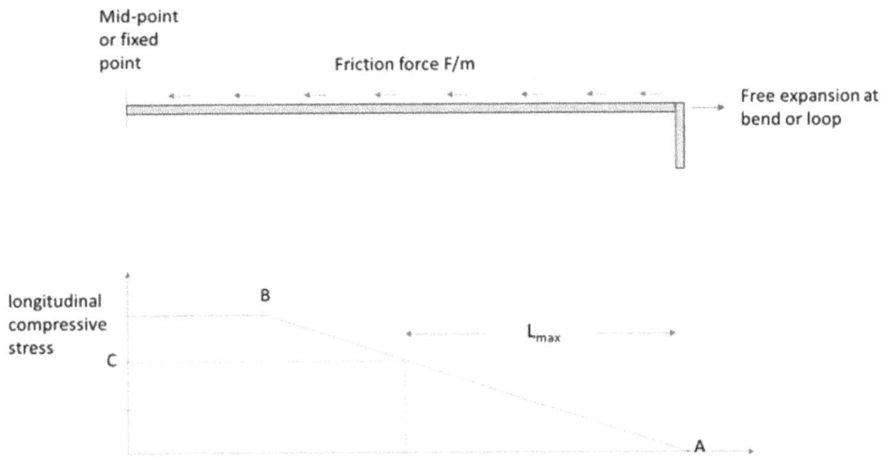

Figure 5.11. Limiting compressive stresses by use of expansion bends or loops

A force balance in the pipe gives:

$$F.\,L = \sigma_c \pi\ dt$$

Where F is assumed to be a constant friction force per m and σ_c is defined as the required limiting stress level, then

$$L_{\max} = \sigma_c \pi\ dt/F, \text{ i. e. } L_{\max} \text{ is proportional to } d.\ t$$

Manufacturer's information provides values of $L_{\max}$ for each pipe diameter.

(B) Pre-heat the pipe before backfilling

In this method the pipe is fully installed in the trench and then heated to a temperature midway between the ambient and the operating temperature. The pipe is allowed to expand in the trench and then backfilled. Then when heated to its operating temperature the thermal expansion compressive stresses will be at a lower more acceptable level. It also means that when cooled down again the pipe will be in tension to a similar level of stress. The pre-heating can be carried out by using hot air, hot water (e.g., the system water) or by electrical means.

This method avoids the use of expansion bends or loops and so can provide some cost savings and potentially less disruption to road traffic. However, it relies on a long length of trench remaining open so that expansion can take place before backfilling which is often not desirable in terms of limiting disruption.

Also it should be noted that although the stresses are then well within the yield stress the stress *range* is the same as for a fully constrained system and it is the stress range that is important when considering risk of low cycle fatigue (see below).

(C) Accepting high levels of compressive stress

As noted above the thermal expansion stresses cannot lead to a structural collapse mode as they are self-limiting. So, some designers consider that the high stresses can be allowed to build up in long straight sections. Sometimes this approach is termed cold-laying. The high compressive stresses carry an increased risk of low cycle fatigue failure and local and gross instability (buckling) discussed below.

5.7.4 Limit state B1—low cycle fatigue

This failure mechanism is low cycle/high stress fatigue which occurs when a material has been subject to large stress cycles multiple times. This would be the case for example where a pipe is fully constrained in the ground and heated and cooled between ambient and its operating temperature as a result of shutdowns for maintenance or from intermittent operation.

EN13941-1 provides assumptions for the *minimum* number of cycles that shall be used in the analysis:

Major pipelines 100
Main pipelines 250
House connections 1000

The Palmen–Miners formula is used to determine whether the design is acceptable:

$$\Sigma^i n_i / N_i < 1/\psi_{fat}$$

where:

n_i is the number of cycles of stress range ΔS_i during the design life
N_i is the number of cycles of stress range ΔS_i to cause failure
ψ_{fat} is the safety factor for fatigue (varies depending on Project Class)
i is the number of different stress ranges

N_i can be calculated from:

$$N_i = (k/S_i)^m = (5000/S_i)^4$$

Annex C of EN13941-1 provides the details necessary to calculate the stress range for different components such as bends and branches together with the stress intensification factors to be applied.

The designer will need to consider the different stress cycles that each part of the network will experience over the lifetime.

For example, two extreme cases can be identified:

Two shutdowns per year over 50 years = 100 stress cycles
Intermittent heating with one shut-down per day over 50 years = 18 250 cycles

Weld quality is important in limiting the risk of failure by low cycle fatigue as any weld will have imperfections which can be the initiation site for crack growth.

The benefits of intermittent operation of a long branch supplying a customer are reduced heat losses. However, with steel pipes there is a risk of failure from fatigue and a controlled bypass flow would normally be used to keep the pipe temperature within reasonable limits.

Even if a bypass is used there could be swings of temperature in the return pipe of say twice a day. So, another case to assess would be:

A reduced stress range but with 36 500 cycles over 50 years.

5.7.5 Other limit states

Limit state B2—high cycle fatigue
This limit state relates to situations where the pipeline may be subject to much more frequent changes of loading, e.g., from vibration or traffic loadings.

Limit state C1—instability—localised buckling
This involves the buckling of the pipe wall when subject to high compressive stress, see figure 5.12. EN13491 shows that this will not occur if the pipe wall is thick enough and pipes supplied to EN253 will meet this criterion.

Limit state C2—instability—gross buckling
If the pipe is considered as a slender column and then subjected to high compressive stress it could buckle. Normally this is prevented by the soil backfill around the pipe even if the compressive stresses reach the yield stress. However, this failure mode could arise if part of the trench has not been backfilled or where subsequent excavation has exposed the pipe when it is still under stress, see figure 5.12.

Limit state D—serviceability limits
This limit state is to cover special situations where there may be external design requirements to be taken into account, e.g., limiting movements at building entries or limiting deflections in the case of pipe bridges.

Limit state for polyurethane foam
Pre-insulated pipes supplied to EN253 are required to demonstrate that the foam maintains its structural integrity and the bond to the steel at continuous

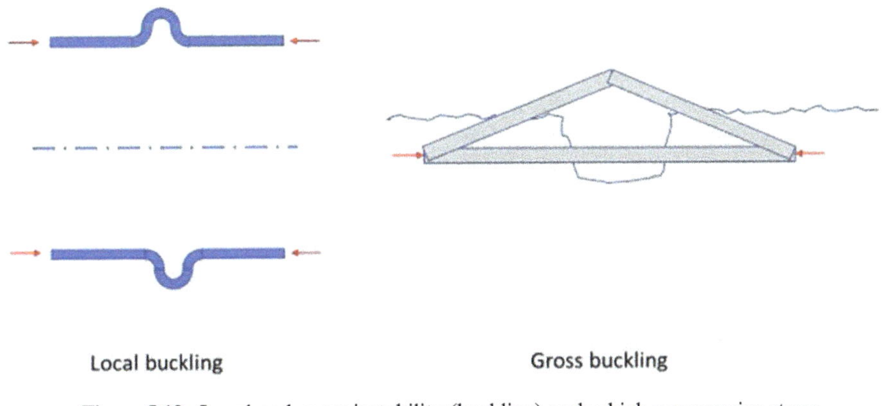

Figure 5.12. Local and gross instability (buckling) under high compressive stress

temperatures of 120 °C and a life of 30 years. For the lower temperatures which are now commonly used the lifetime will be significantly more than 30 years.

Limit state for PE casing
EN13941-1 requires that the PE casing is not subject to temperatures more than 50 ° C. A check should be carried out where foam cushions are used for expansion and where high system temperatures are used.

5.7.6 Software available

To assist the designer, software is available to carry out the necessary calculations in accordance with EN13941. This work is often carried out by the supplier of the pipe system. An example is the sisKMR software developed by the German company GEF Ingenieur AG [25].

For a more detailed description of the stress analysis of pre-insulated pipe systems see [26].

5.8 Quality control in construction

The nature of a DH/C system is that it is buried in the ground and designed to be maintenance free. As a result, the quality of the initial installation is paramount. Also, if there are faults in the installation most of these will not show up as a problem for many years to come, e.g. ingress of ground water through joints causing external corrosion is a failure mechanism that make take many years to result in a noticeable leakage of system water. Failure from low cycle fatigue which may be initiated at a weld defect similarly may take many years before causing a failure.

Quality control is therefore important and CIBSE CP1.2 [4] requires as a minimum the following steps in the installation to be subject to checks which need to be documented:

- Inspection of the trench prior to installing pipes
- NDT of welds and tightness tests
- Joint closure welding or shrinking
- Joint closure tightness test (e.g., air pressure test)
- Continuity test for surveillance system
- Trench inspection prior to backfilling
- Compaction around pipes and laying of marker tape
- Flushing and initial filling of pipe
- System pressure test (strength test)
- Backfilling and surface reinstatement

EN13941 Part 2 details the minimum requirements for NDT testing (depending on Project Class) and the acceptable methods for tightness testing of welds including air pressure tests.

For Project Class A the testing requirements can be summarised as:

- Tightness test using air at 0.2 bar g and soapy liquid around weld

- NDT using x-ray or ultrasonics to 10% of welds
- Optional strength test for entire system at 1.5 times working pressure

For plastic pipes the weld NDT and surveillance system tests will not be relevant but appropriate checks on carrier pipe joints should be incorporated in the quality control regime.

5.9 Isolating valves

Maintenance of thermal substations/building interfaces is assumed to be possible as typically there would be an isolation valve as part of the substation.

If maintenance is required on the DH/C network then isolation of parts of the circuit will mean that fewer customers are impacted. For example, if there is fault in the branch supplying a single customer then if the branch has isolating valves then only this customer would be affected. Installing isolation valves within the main lines of the network would allow upstream customers to be supplied even if there was a downstream fault.

Isolating valves will, however, add to the cost of the network and introduce a possible source of weakness and additional maintenance requirements, so a judgement has to be made as to their number and location.

Conventional valve chambers have been superseded by pre-insulated ball valve assemblies which can be buried in the ground with access only required to operate the valve spindle. The relevant standard is EN488 [27]. It is important that isolating valve chambers are designed to be free draining so that the risk of corrosion of the spindle mechanism is reduced, although stainless steel spindles should be used. An annual inspection and trial operation of the valve is normally recommended.

Drain and venting points are often incorporated at valve positions to assist in draining down and refilling during maintenance work.

5.10 Design for future growth of demand

In the UK it has been necessary to market DH in competition to the established market of gas-fired boilers and take up by customers is not usually mandated. So it is necessary to plan a network to supply an initial group of customers (often called anchor customers and typically public sector buildings) but with the potential to supply other customers in the area over time. This has implications for the sizing of pipes to allow for this expansion.

As described in section 5.5.5 an oversized pipe will result in a higher capex but a lower pumping cost than the optimum. And if the pipe is undersized for the final load the pumping energy would be higher but heat losses and capex would be lower than optimal. The design of a DH network has considerable flexibility with up to 50% increase in demand being able to be incorporated without significant overall cost penalty, subject to total pressure constraints.

Other factors to take into account in planning a network are:

- Potential for new buildings in the area (densification); these may increase the floor area to be connected but the higher thermal energy standards of new buildings need to be taken into account.
- The risk of demolition of existing buildings or the relocation of occupants and subsequent redevelopment causing a temporary reduction in the supply of heat.

Discussions with the local authority planning department will clearly be important to understand the broader development aspirations for the area.

5.11 District cooling pipes

Pre-insulated pipes complying with EN253 [5] and associated standards are also widely used for District Cooling so much of the information in this chapter is relevant for DC as well as DH. When insulating chilled water pipes there is a need for a vapour barrier to prevent condensation on the pipe which with steel pipe could lead to corrosion. The use of the pre-insulated piping systems to EN253 which is designed to prevent ground water ingress will perform the same function in preventing condensation. Nevertheless, in some climates (hot and humid) the water vapour in the air will become trapped in the joint area and pre-cooling of the joint before sealing is recommended.

The risk of corrosion from condensation is avoided if plastic or composite carrier pipes are used. These will have a very long life as the operating temperatures are low.

The main differences between DC and DH are:

- The small ΔT available on the DC network means that pipe sizes are larger and so it is even more important to minimise network lengths and optimise pipe sizes.
- As described in section 5.2 the pumping energy also needs to be minimised as this contributes to unwanted heat gain in the circuit.
- Although the stress calculations for steel pipe are the same in principle the impact of thermal expansion/contraction is much less and so no provision for expansion is necessary, although building entries still need to provide for pipe movement so that forces are not transmitted to the building structure.

Bibliography

[1] Health and Safety Executive *Avoiding Danger from Underground Services HSG47* (3rd edn) (https://hse.gov.uk/pubns/books/hsg47.htm) (hse.gov.uk)

[2] BSI PAS128:2022 *Underground Utility Detection* (British Standards Institution) (https://bsigroup.com/en-GB/standards/pas-1282022/ PAS 128:2022)

[3] *New Roads and Street Works Act 1991* (https://legislation.gov.uk/ukpga/1991/22/contents) (legislation.gov.uk)

[4] CIBSE 2020 *Heat Networks, Code of Practice for the UK, CP1.2* (CIBSE)

[5] EN253:2019 2019 *District Heating Pipes. Bonded Single Pipe Systems for Directly Buried Hot Water Networks. Factory Made Pipe Assembly of Steel Service Pipe, Polyurethane Thermal Insulation and a Casing of Polyethylene* (European Committee for Standardization)

[6] Massey B and John W-S 2006 *Mechanics of Fluids* 8th edn (London: Taylor and Francis) p 248

[7] CIBSE 2018 *Guide C—Reference Data* (CIBSE)

[8] Massey B and John W-S 2006 *Mechanics of Fluids* 8th edn (London: Taylor and Francis) p 254

[9] Martin-Du Pan O, Woods P and Hanson-Graville R 2019 Optimising pipe sizing and operating temperatures for district heating networks to minimise operational energy consumption *Build. Serv. Eng. Res. Technol.* **40** 237–55

[10] Fluidit Oy *Fluidit Heat* (https://fluidit.com/software/fluidit-heat/)

[11] Schneider Electric *Termis Engineering* (https://se.com/il/en/product-range/61613-termis-engineering/#overview)

[12] Vitec Energy Netsim grid simulation *Netsim* (https://vitec-energy.com/netsim-grid-simulation/)

[13] EN489-1:2019 *District Heating Pipes. Bonded Single and Twin Pipe Systems for Buried Hot Water Networks—Joint Casing Assemblies and Thermal Insulation for Hot Water Networks in Accordance with EN 13941-1* (European Committee for Standardization)

[14] EN448:2019 2019 *District Heating Pipes. Bonded Single Pipe Systems for Directly Buried Hot Water Networks. Factory Made Fitting Assemblies of Steel Service Pipes, Polyurethane Thermal Insulation and a Casing of Polyethylene* (European Committee for Standardization)

[15] EN14419:2019 2019 *District Heating Pipes. Bonded Single and Twin Pipe Systems for Buried Hot Water Networks. Surveillance Systems* (European Committee for Standardization)

[16] EN15632-1:2022 2022 *District Heating Pipes. Factory Made Flexible Pipe Systems—Classification, General Requirements and Test Methods* (European Committee for Standardization)

[17] EN15632–2:2022 2022 *District Heating Pipes. Factory Made Flexible Pipe Systems—Bonded System with Plastic Service Pipes; Requirements and Test Methods* (European Committee for Standardization)

[18] EN15632–3:2022 2022 *District Heating Pipes. Factory Made Flexible Pipe Systems—Non Bonded System with Plastic Service Pipes; Requirements and Test Methods* (European Committee for Standardization)

[19] EN ISO 13670 1998 *Plastics Pipes for the Conveyance of Fluids Under Pressure. Miners Rule. Calculation Method for Cumulative Damage* (International Organisation for Standarization)

[20] EN15632-4:2022 2022 *District Heating Pipes. Factory Made Flexible Pipe Systems—Bonded System with Metal Service Pipes; Requirements and Test Methods* (European Committee for Standardization)

[21] EN15698-1:2019 2019 *District Heating Pipes. Bonded Twin Pipe Systems for Directly Buried Hot Water Networks. Factory Made Twin Pipe Assembly of Steel Service Pipes, Polyurethane Thermal Insulation and One Casing of Polyethylene* (European Committee for Standardization)

[22] EN15698-2:2019 2019 *District Heating Pipes. Bonded Twin Pipe Systems for Directly Buried Hot Water Networks. Factory Made Fitting and Valve Assemblies of Steel Service Pipes, Polyurethane Thermal Insulation and One Casing of Polyethylene* (European Committee for Standardization)

[23] EN13941-1:2019 + A1 2021 *District Heating Pipes. Design and Installation of Thermal Insulated Bonded Single and Twin Pipe Systems for Directly Buried Hot Water Networks—Part 1: Design* (European Committee for Standardization)

[24] EN13941-2 2019 (main) 2019 *District Heating Pipes. Design and Installation of Thermal Insulated Bonded Single and Twin Pipe Systems for Directly Buried Hot Water Networks—Part 2: Installation* (European Committee for Standardization)

[25] GEF 2023 *GEF Ingenieur AG. sisKMR Software* (https://siskmr.com/home/sisKMR—Software für statische Berechnungen)

[26] Frederiksen S and Sven W 2013 *District Heating and Cooling* (Studentlitteratur) ch 8

[27] EN488:2019 2019 *District Heating Pipes. Bonded Single Pipe Systems for Directly Buried Hot Water Networks. Factory Made Steel Valve Assembly for Steel Service Pipes, Polyurethane Thermal Insulation and a Casing of Polyethylene* (European Committee for Standardization)

Chapter 6

Heating and cooling production

6.1 Introduction

DH and DC are technologies for transporting thermal energy in the form of hot water or chilled water from a production plant to the customer. Whether this is advantageous in economic or environmental terms depends on the effectiveness of the production plant.

The production plant needs to produce thermal energy more efficiently and at significantly lower cost than the systems that would otherwise be installed in buildings so as to compensate for the cost of the DH or DC network, its thermal losses/gains and the pumping energy.

The DH/C production plant typically uses a wider range of technologies than possible in individual buildings. It can use combinations of technologies which would be too complex to install in an individual building. It is also easier to incorporate thermal storage at the production plant than in individual buildings where there are likely to be more space constraints.

The key technologies that produce heat for DH can be classified as follows:

Technologies which use more efficient processes:

- Combined heat and power (CHP)—which is more efficient than separate production of heat and electricity (when using the same fuel)
- Heat pumps using ambient sources (air, ground and water)—which is more efficient than direct electric heating

High temperature heat recovery from process plant

- Energy from waste plants
- Thermal power stations
- Industrial heat recovery at high temperature

Low temperature heat recovery with heat pumps

- Datacentres
- Mine water
- Industry

doi:10.1088/978-0-7503-5286-4ch6

- Power stations
- Waste water and sewage

Renewable heat sources

- Biomass
- Deep geothermal
- Solar

These low carbon primary sources will typically meet 80%–90% of the heat demand. In addition, there are likely to be peaking boilers for the remaining demand, standby boilers for security of supply and thermal stores. These elements may be sited at the production plant or within the network.

The key technologies that are used in the production of coolth for DC are:

- Compression chillers
- Heat driven chillers
- Free cooling associated with heat pumps
- Free cooling associated with water sources (ice melt in rivers)
- Evaporative cooling towers (more efficient than using air-cooled chillers)
- Coolth storage

This chapter discusses the engineering of these technologies including how they might be combined. More detail on individual technologies can be found in the specialist sources for each technology listed in the Bibliography.

Chapter 10 discusses technologies that may become more important in the future including hydrogen, small modular nuclear stations and carbon capture and storage.

6.2 Combined heat and power—a more efficient process

6.2.1 Principles

Combined heat and power (CHP) is defined as the simultaneous production of heat and electricity in a single process. It can take many forms with a range of scales and fuels. It has often been the main source of heat for DH as it has produced heat at low cost and reduced CO_2 emissions. So much so that the UK's District Heating Association was renamed the CHP Association in the 1980s recognising that if the challenges associated with CHP could be solved (e.g., fair access to electricity markets) then DH would be able to expand more rapidly. Subsequently, the CHPA changed its name to Association for Decentralised Energy to reflect better the range of technologies that now supply DH/C schemes.

In the past, gas-fired CHP was of benefit because it displaced both coal and gas power stations so CO_2 reductions were obtained from both the efficiency of CHP and fuel switching from coal to gas.

In recent years, CHP has been displacing gas-fired power stations but in the future, as the grid decarbonises, the benefits of CHP will decline further as the hours in the year when its operation will lead to CO_2 savings will be curtailed. As a result,

Government policies such as the Green Heat Network Fund are not supporting small-scale gas-fired CHP.

Nevertheless, a number of existing DH schemes use gas-fired CHP as their primary heat source so it is important to understand this technology and how to assess its benefits.

Thermodynamic fundamentals

The First Law of Thermodynamics states that energy is conserved, i.e., it cannot be created or destroyed within any defined system boundary.

We are most familiar with this in the case of a boiler in one's home. The efficiency of the boiler is defined as:

$$\eta_b = \text{useful heat/fuel input}$$

With CHP, electricity is also produced and the overall energy balance is maintained as:

$$\text{Heat input} = \text{useful heat} + \text{electricity} + \text{losses}$$

To define the efficiency of a CHP plant two numbers are required. These can be any two of:

$$\text{Electrical efficiency} = \text{electricity/fuel input}$$
$$\text{Thermal efficiency} = \text{useful heat/fuel input}$$
$$\text{Total efficiency} = (\text{electricity} + \text{useful heat})/\text{fuel input}$$

Electricity is produced when a generator is turned by a shaft, i.e., it is mechanical work which is then used to create electricity with some losses in the generator

The First Law of Thermodynamics states that energy is conserved and makes no distinction between heat and work.

The Second Law of Thermodynamics *does* distinguish between heat and work. It states that work can be turned into heat but heat can only be turned into work if some heat is transferred to a second reservoir as illustrated in figure 6.1.

The efficiency of the conversion from heat to work is governed by the absolute temperatures of the heat source (T_1) and the second reservoir where heat is rejected (T_2). In an ideal world of perfect equipment the work efficiency is given by:

$$\mathrm{E_w} = W/Q_1 = (T_1 - T_2)/T_1 \tag{6.1}$$

(see standard thermodynamic textbooks including [1]).

From this equation, to maximise efficiency, the heat source temperature needs to be as high as possible and the second heat reservoir temperature needs to be as low as possible.

If the heat rejected to the second reservoir is as close to ambient temperature as possible then it will not be useful for heating purposes. A CHP plant operates with this temperature raised to a more useful level and the work efficiency falls in accordance with equation (6.1).

As a result, although the overall system is more efficient, the heat is not 'free' or without environmental impact as the efficiency of generating work is reduced.

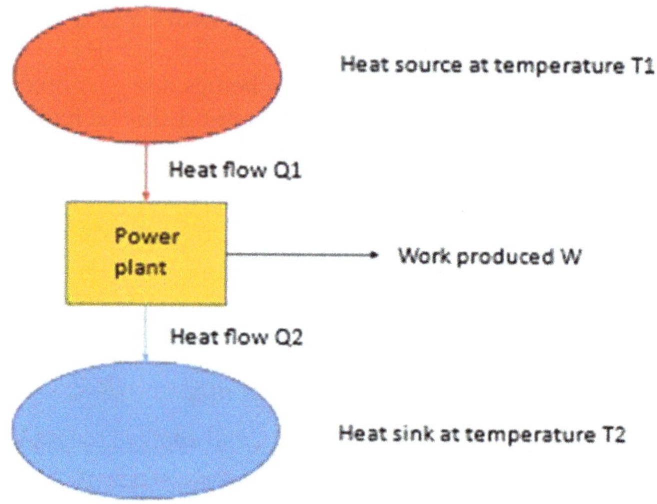

Figure 6.1. Thermodynamic principles for thermal power electricity generation.

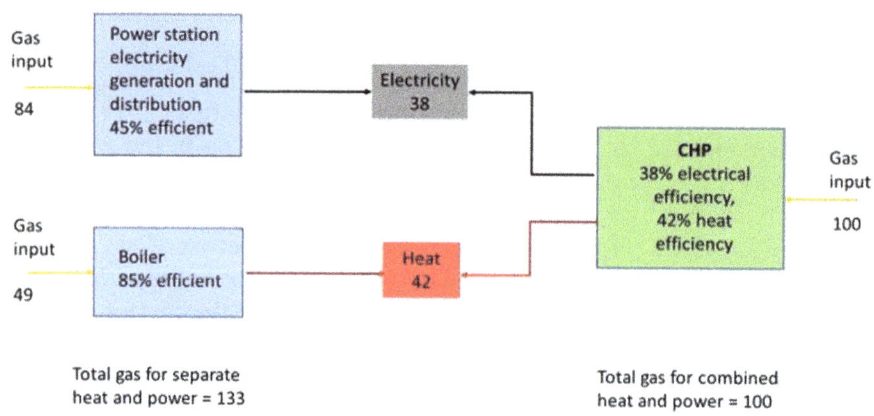

Figure 6.2. Energy savings from operation of CHP plant.

Figure 6.2 compares generating heat and power separately with CHP and it can be seen that although the electrical (work) efficiency is lower with CHP there is an overall improvement of efficiency of the fuel use, reducing by 25% (133 units down to 100). If this saving is allocated to heat then the fuel use for heating reduces from 49 to a net fuel use for CHP of 100 – 84 = 16 or a 67% saving. It is this improvement that justifies the use of CHP and often the economic basis for DH.

6.2.2 Types of CHP plant

Spark-ignition gas-engine CHP

Although compression ignition dual-fuel engines have been used for CHP, the need for pilot oil to initiate combustion of natural gas means that their CO_2 emissions are higher and not compensated for by their higher electrical efficiency. Spark-ignition gas-engines (SIGE) therefore predominate for gas-fuelled CHP in the range 500 kWe to 5 MWe for temperatures <100 °C that are typically used to supply DH. Multiple units can be installed to supply larger schemes.

Heat recovery from a SIGE is ideally suited to DH applications as heat can be recovered at temperatures suiting modern DH. Heat recovery is possible from:

- Exhaust gas—this leaves the engine at about 450 °C and is typically cooled to 120 °C
- Engine jacket and oil cooler—this typically operates at about 90 °C
- Intercooler—the turbo charger increases the pressure of the combustion air which also raises its temperature; the engine is more efficient if this warm air is then cooled to around 40 °C

A typical energy balance for a large (3 MWe) SIGE is given in figure 6.3.

The main disadvantage of a SIGE is the local NOx emissions which can impact local air quality especially in inner city areas. It is now often a requirement to reduce emissions using selective catalytic reduction although this adds to the initial cost and requires additional space.

SIGE systems are used with both natural gas and biogas with a high methane content. Recent developments include the use of hydrogen blends and pure hydrogen.

Where natural gas is used as a fuel, the CO_2 savings depend on the CO_2 displaced by the electricity generated by CHP. A net saving is only realised if the electricity displaced is also being generated by fossil fuels. This is discussed further in section 6.4.

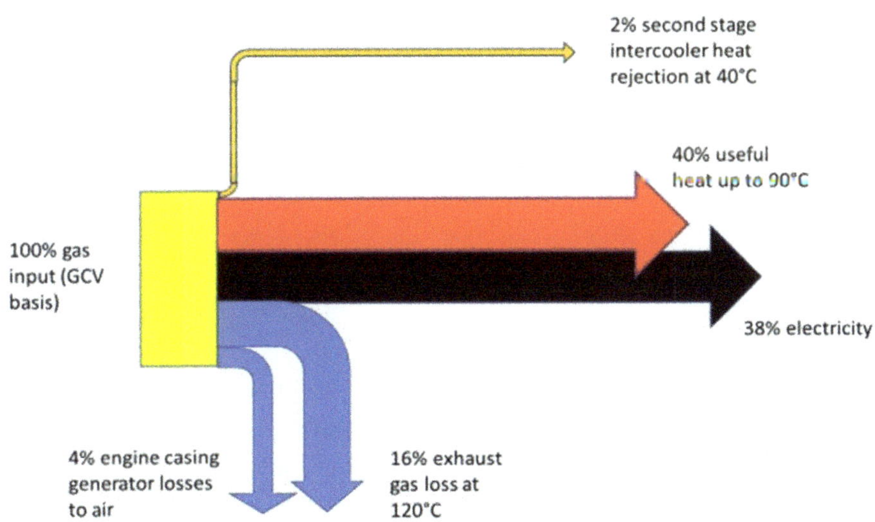

Figure 6.3. Energy balance for a typical large spark-ignition gas-engine CHP.

Gas turbine CHP

Open cycle gas turbine (OCGT) CHP is also available for a wide range of sizes however it will have a lower electrical efficiency than a SIGE (especially if gas compression is required). Its main advantage is providing heat at a higher temperature (suitable for steam raising) and so has been used mainly for industrial CHP applications rather than DH where high temperatures are not required.

Combined cycle gas turbine (CCGT) plant uses the heat from the gas turbine to raise steam which then drives a steam turbine. The complexity of this type of plant means that the extra cost is more justified if it is at a larger-scale and operating as a base-load power station with heat extracted for DH as required. This type of CHP is discussed in section 6.5.

Fuel cell CHP

Fuel cells are a non-combustion alternative for small-scale CHP and, when used with natural gas, will consist of a reformer stage to convert methane to hydrogen and CO_2 followed by the fuel cell which uses hydrogen and oxygen to produce electricity and water, releasing heat in the process. A fuel cell is categorised by the type of electrolyte used.

Fuel cells potentially have the advantages of:

- Slightly higher electrical efficiency
- Very low emissions to air
- CO_2 capture and storage

Although well proven in certain applications, e.g., space and automotive they have yet to become established for stationary applications. The disadvantages are: high initial cost, limited lifetime of components and the need to operate the fuel cell relatively continuously.

For further information on the design of CHP systems see CIBSE AM12 [2].

Box 6.1. King's Cross redevelopment, London.

Project Name: King's Cross Heat Network, London

DH Developer: Metropolitan King's Cross

Construction Contractor: Vital Energi

Operator: Metropolitan Kings Cross

This King's Cross Central 2019.jpg image has been obtained by the author(s) from the Wikimedia website where it was made available under a CC BY-SA 4.0 licence. It is included within this book on that basis. It is attributed to Crookesmoor.

Key data:

Peak design heat demand (MW):	45
Heat sources:	Gas-fired CHP, gas-fired boilers
Date construction commenced:	2008

Description

The King's Cross development is a major mixed-use development on a brownfield site adjacent to King's Cross station. Development started in 2008 and the last buildings are now in construction. Customers include: residential, commercial offices, retail space and educational buildings.

The DH network was installed from the outset initially using a temporary boiler plant and later an Energy Centre was built in the basement of one of the new buildings. This comprises two gas-fired spark-ignition engine CHPs each of 2 MWe, 2.1 MWth and 33 MW of gas-fired boilers for peak and standby use. There are also two 74 m^3 thermal stores connected in series.

The heat network consists mainly of flow and return pre-insulated steel pipes. The main challenge in the construction phase was expanding the scheme gradually within a major construction site whilst maintaining supplies to buildings already connected.

The system operates with a flow temperature of 90 °C maintained through the year. Return temperatures are currently around 65 °C.

Plans for the future include the progressive decarbonisation of the network via installation of heat pumps.

Information kindly supplied by Metropolitan

6.3 Heat pumps—a more efficient use of electricity

6.3.1 Principles

As the grid progressively moves towards a renewable and low carbon supply the production of heat from heat pumps will become more viable and ultimately replace the use of fossil fueled CHP.

Fundamental thermodynamics

The First Law analysis of a heat pump shows that:

$$\text{Useful heat} = \text{work input} + \text{heat from lower temperature source}$$

The Second Law states that work will always be required if heat is to be transferred from a lower temperature body to a higher temperature, as illustrated in figure 6.4.

The efficiency of the heat pump is known as the coefficient of performance (CoP) and is governed by the absolute temperatures of the heat delivered (T_1) and the heat source (T_2). In an ideal world of perfect equipment:

$$\begin{aligned} \text{CoP} &= \text{useful heat/work input} = Q_1/W \\ &= T_1/(T_1 - T_2) \end{aligned} \tag{6.2}$$

(see standard thermodynamic textbooks including [3]).

Hence efficiency is improved if the difference between T_1 and T_2 is reduced. This means that a more efficient heat pump system will be realised when the heat source is at as high a temperature as possible and the useful heat is produced at as low a temperature as possible. Figure 6.5 shows how the CoP will vary with supply temperature for three heat source temperatures taking a typical Carnot efficiency of 50% (the factor applied to equation (6.2)).

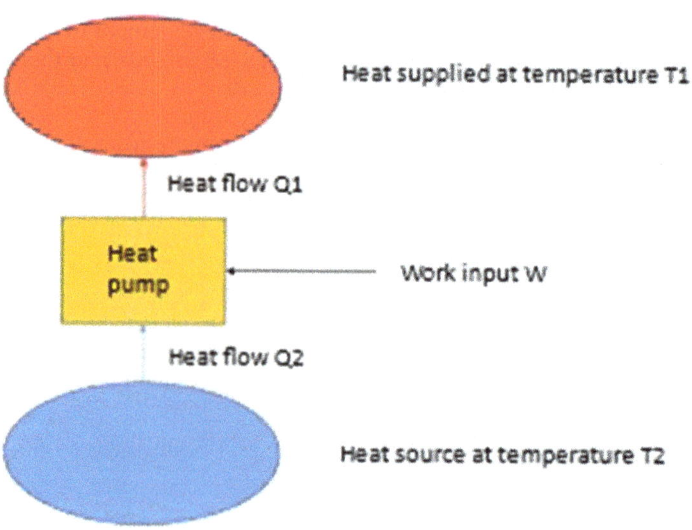

Figure 6.4. Thermodynamic principles for heat pump.

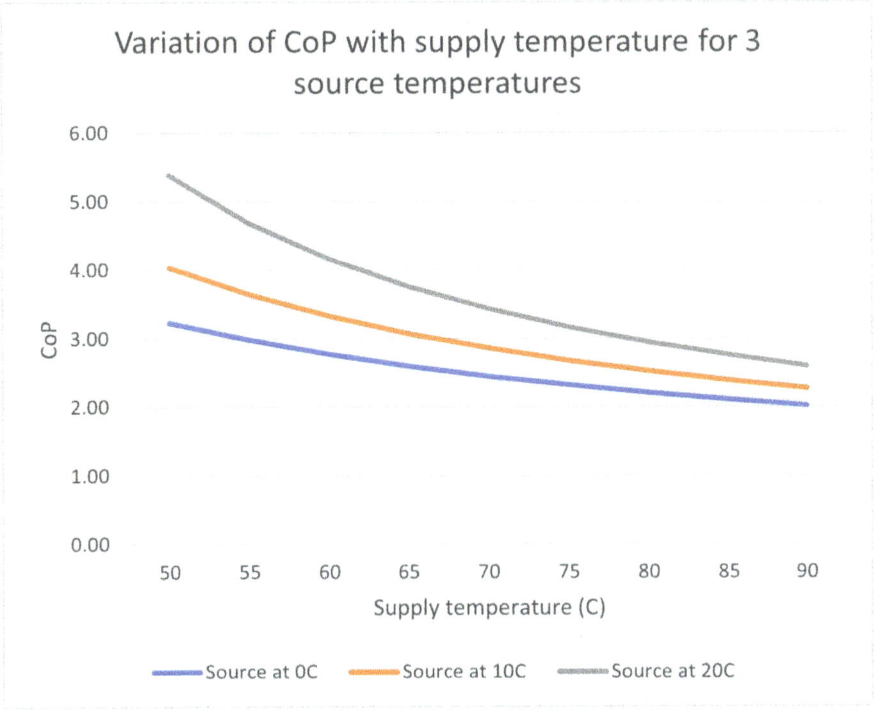

Figure 6.5. Variation in CoP with supply temperature for three source temps, Carnot efficiency of 50% assumed.

It can be seen that there are efficiency advantages in designing a DH scheme for low temperature operation say 60 °C as is now commonly achieved especially with new-build projects (https://www.gtc-uk.co.uk/news/gtc-reveals-a-low-carbon-heating-solution-to-replace-the-use-of-gas-for-new-homes/)

A DH system may have a relatively wide temperature difference between flow and return. This means that arranging two or more heat pumps in series will in principle bring efficiency benefits as it is only the final heat pump that will need to deliver heat at the required flow temperature and the others can deliver heat at intermediate temperatures at higher CoPs. However, the choice of refrigerant and how well matched the heat pump performance is to the required flow and return temperatures is a more significant issue.

6.3.2 Types of heat pump

Air source heat pump

An air source heat pump uses ambient air as the heat source. It has the benefit of being able to be sited anywhere (the heat source is evenly distributed). The disadvantage is that in winter the air temperature will be lower and more variable than other ambient sources (in summer the reverse is true but heat demand is higher in winter). Also, the heat exchanger between the air and the heat pump circuits will be prone to freezing as humidity in the air turns to ice which will reduce the

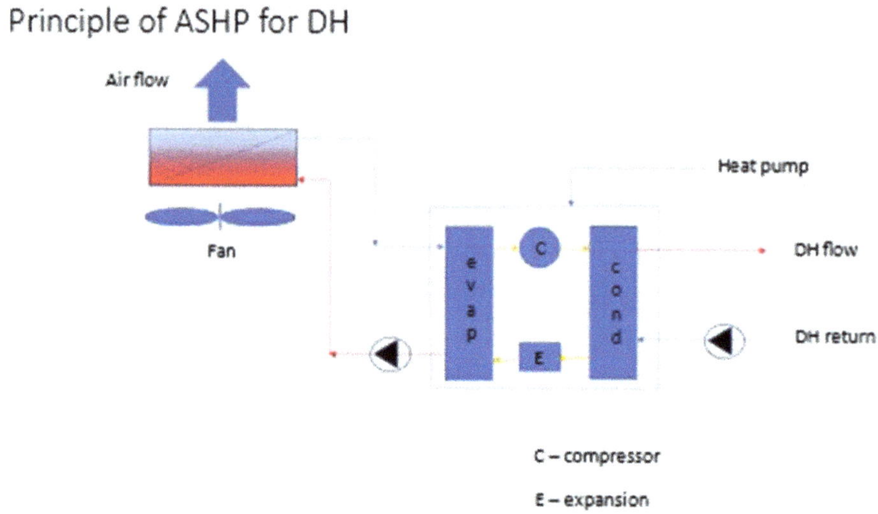

Figure 6.6. Diagram to show principles of an ASHP system with intermediate water circuit.

effectiveness of the heat exchanger. This in turn will require heat to be used in a defrost cycle to clear the ice so that heat transfer rates can be maintained. Suitable drainage must be provided to clear the water produced in the defrost cycle without causing a hazard if it refreezes.

The ASHP can either circulate the refrigerant through the heat exchanger with the air or the refrigerant can connect to a separate brine circuit which then exchanges heat with the air (see figure 6.6). The former is preferable for overall efficiency as it avoids an additional stage of heat transfer and the associated temperature difference. The latter has the advantage of minimising the amount of refrigerant and keeping this in a controlled area within an energy centre.

Heat transfer from air to liquid requires significant surface area for the heat exchanger and the air to be moved at high velocity using a fan. This, therefore, involves significant fan energy which can result in adverse noise impacts in built-up areas.

The design needs to ensure that the cooled air from the heat pump is dissipated and there is no significant recirculation otherwise performance will be affected. Also ‘cold pluming’ where cold air drops to ground level causing annoyance should be evaluated.

Water source heat pump

Water source heat pumps may use ground water, lake or river water, or sea water as a heat source. The advantage is that the temperature of such water sources vary less than air temperature. With large bodies of water there may be a lag between the change in air temperature and the change in water temperature. The sea temperature is an example where in winter the sea temperature is a few degrees warmer than the air temperature and the maximum temperature occurs one to two months after the maximum air temperature (see figure 6.7). However, the differences are relatively small and any benefit can be outweighed by other factors in the cost and energy requirements of the system.

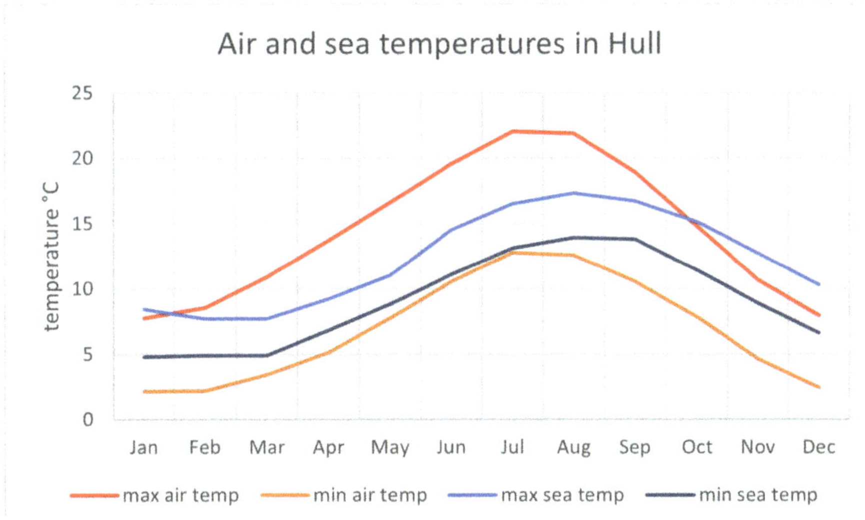

Figure 6.7. Diagram of sea temperature vs air temperature in Hull, UK.

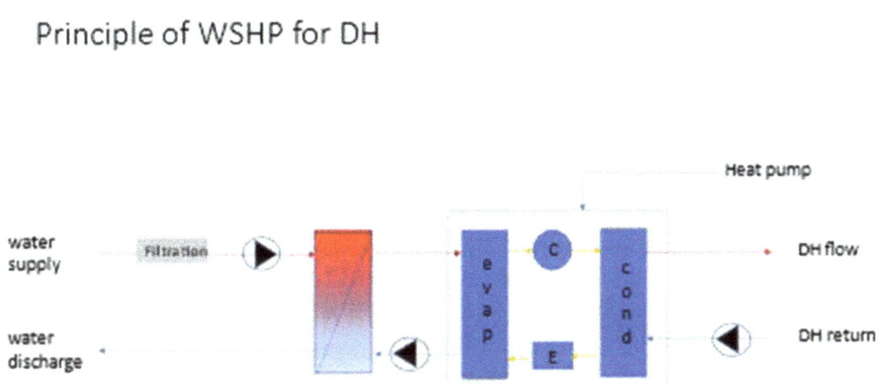

Figure 6.8. Diagram to show principles of a WSHP system.

Smaller water bodies and small rivers will tend to have temperatures which track the average daily air temperature more closely and so the performance advantage over ASHP will be less.

Ancillary energy is needed to access the water source, e.g., pumping from the ground a lake or the sea and this may involve a significant change of height. Close attention needs to be given to the need for filtration of the water extracted (see figure 6.8). However, heat transfer from water to refrigerant is more effective than with air and so a smaller temperature difference is needed and fan power is avoided.

The main disadvantage is that the production plant will need to be fairly close to the water source which may be a constraint given that such locations often have higher value land and are often in demand for housing.

River water may not be available throughout the year, i.e., limited volumes may be present in summer.

Generally, extracting water from a river and returning it a few degrees lower in temperature will not be seen as damaging to the local ecosystem and may in fact be beneficial.

Ground water is attractive as the temperature will be very stable through the year, however, the amount of water that can be extracted from a given borehole is uncertain and even a trial borehole is expensive.

Ground source heat pump

The ground itself can also be used as a heat source using closed loop boreholes or horizontal helical tubes. The heat transfer between the ground and an embedded pipe varies with the type of soil, the water content and the temperature difference between the ground and the circulating fluid, however figures of 40–50 W m^{-1} are typical. A thermal response test is recommended using a trial borehole that can be incorporated into the final scheme. A relatively large area is required to achieve the necessary heat transfer even using vertical boreholes. Whilst this may be feasible for small schemes, e.g., a house with a large garden or a small group of houses, for a DH scheme which involves centralising heat production the area required is unlikely to be available.

It may be possible to combine air source, ground and water source heat pumps to optimise the resources and land available.

6.3.3 Choice of refrigerants for heat pumps

An important potential benefit in using DH is that the centralised heat pumps can use a wider range of refrigerants than are typically used in individual building heat pump applications. This includes refrigerants where the impact on global warming from leakage is zero or very low. Such refrigerants may have higher risks associated with toxicity and flammability but these can be managed in a large central energy centre more easily than in individual buildings.

The thermodynamic behaviour and properties of fluids will mean that the optimal operating range for each can be very different. The main considerations in selection of a refrigerant for a particular application are:

- The source and sink temperature requirements—this will include both the temperature requirements and the temperature differentials.
- Global warming potential—all refrigerant systems will be subject to leaks and some refrigerants have global warming potential (GWP) many times that of CO_2; this can lead to otherwise low carbon systems having a significantly higher overall and embodied carbon emissions.
- Toxicity—many refrigerants are toxic and consideration should be given to the ability to control any discharges such that hazards to health and the environment are avoided.
- Flammability—some refrigerants, particularly ones based on hydrocarbons, can be highly flammable and control of risks related to leakages and fire/explosion risks need to be managed.

As the electricity supply is decarbonised the relative importance of refrigerant leakage will grow. So, it is possible that a DH or DC system using a zero global warming refrigerant such as ammonia could have a lower environmental impact even if it was less efficient in energy terms than an individual building solution using a refrigerant with a higher GWP.

The production of HFC refrigerants is being phased out according to the Kigali agreement so, even if the initial heat pump is selected using this type, consideration needs to be given to the needs during operation and an eventual replacement with alternative. This process is increasing the costs of refrigerants needed for future recharging.

The global warming potential (GWP) index compares the impact of refrigerants in relation to CO_2. Guidance on leakage rates is given in CIBSE TM65 [4]. Refrigerants have differing safety implications for flammability and toxicity. For further details see ISO 817 [5]. Table 6.1 compares refrigerants that are commonly used for heat pumps.

For further information on the design of heat pumps see the following:

For surface water source heat pumps CIBSE CP2 [6]

For open-loop ground water source heat pumps CIBSE CP3 [7]

For multi-residential buildings CIBSE AM16 [8]

For non-residential buildings CIBSE AM17 [9]

Table 6.1. Comparison of typical refrigerants.

Refrigerant	Global warming potential	Flammability	Toxicity	Comments
R32	675	Mildly flammable	Low	Production is being restricted according to the Kigali agreement
R134a	1430	Non-flammable	Low	Production is being restricted according to the Kigali agreement
Ammonia (R717)	0	Mildly flammable	High	Use is well established in industrial applications so risks are understood. Higher efficiency in high temperature applications
Isobutane (R600a)	3	High	Low	
Propane (R290)	3	High	Low	
CO_2 (R744)	1	Non-flammable	Low (but asphyxiation risk)	Requires return temps <40 °C Higher efficiency in high temperature/high differential temperature applications
HFO 1234ze	7	Mildly flammable	Low	

Box 6.2. Old Market Heat Network, Bristol.

Project Name: Castle Park Harbour water heat pump

Local Authority: Bristol City Council

Developer: Bristol City Council

Design and Build Contractor: Vital Energi

Operator: Vattenfall

Lead consultant: Sustainable Energy Ltd

Key data

Peak heat demand (MW):	6 in 2023
Annual heat delivered to customers (MWh):	6313 in 2022/23
Heat sources:	Water source heat pump, gas-fired boilers
Date construction commenced:	2019
Date first heat delivered:	March 22

Description

The Old Market heat network has been developed by Bristol City Council from 2018 until 2023. The customers of the scheme are a mix of residential and commercial with eight buildings currently supplied and circa 700 dwellings.

The energy centre comprises a 3 MW water source heat pump using ammonia as refrigerant and 3 MW of gas-fired boilers for peak and standby use. There is also a 90 m^3 thermal store. Water is drawn from the harbour and filtered before being passed to the heat pump.

The heat network consists mainly of flow and return pre-insulated steel pipes. The main challenge in the construction phase was crossing historic castle walls and culverts.

The system operates with a peak flow temperature of 80 °C which is reduced to 70 °C in summer. Return temperatures are currently around 55 °C in summer and 50 °C in winter.

Plans for the future include the extension of the network to connect with the other city centre networks and densification of more connections along the existing spine network.

This project is part of the Bristol City Leap initiative and is now operated by Vattenfall as part of the portfolio of heat networks supplying Bristol.

Information kindly supplied by Sustainable Energy Ltd

6.4 Comparison of energy and environmental impact of boilers, CHP and heat pumps

6.4.1 Marginal plant assumption

CHP in the form of a natural gas fuelled gas-engine only provides environmental benefits if it is displacing electricity produced from fossil fuels. When CHP displaces electricity produced from renewable electricity—solar and wind for example then CHP will increase CO_2 emissions to about twice that of a boiler.

In this section, CHP, heat pumps and boilers are compared assuming that natural gas is the primary fuel used for both the national electricity system and for the CHP plant. If, for a given project, CHP is selected the electricity demand will reduce and if a heat pump is selected the electricity demand will increase. These small changes in demand will only impact the gas-fired power stations as wind, solar and nuclear have low operating costs and will operate at their maximum output regardless of the level of demand. The gas-fired power stations are referred to as the marginal plant. In 2021 gas-fired power stations supplied 40% of the total demand, from the Digest of UK Energy Statistics (DUKES) [11] and there were no periods when gas-fired power stations were not required to operate.

The methodology presented below which compares heat supply options by assessing the impact on the marginal gas-fired plant is, therefore, justified for present conditions. The impact of increasing levels of renewable energy in the longer-term on these calculations is discussed in sections 6.9, 9.2 and 10.2.

Two parameters are used for the comparisons presented below:

Primary energy efficiency (and the reciprocal—the Fuel Use Rate)

CO_2 emissions per unit of heat

6.4.2 Primary energy efficiency

Boilers

The concept of the efficiency of a boiler is a familiar one:

$$\text{Efficiency} = \text{heat out/fuel in}$$

Typically, an efficiency of 80% is assumed. For domestic boilers the test bed efficiency as quoted by SEDBUK is higher but this is based on standard operating conditions that assume condensing operation and only applies to the space heating circuit. For combi boilers the efficiency of domestic water heating was found to be lower at around 75% [10].

Efficiency of grid electricity supply

For both CHP and heat pumps it is first necessary to establish the efficiency of the grid supplied electricity at the site in question.

As fossil fuel electricity is almost all obtained from gas-fired combined cycle gas turbine (CCGT) power stations, data has been taken from the DUKES [11] to determine the efficiency of electricity generation.

Table 5.6 of DUKES provides data for the year 2021:
For gas thermal power stations:
Fuel used: 228 242 GWh
Generation: 110 681 GWh
Used on works: 1910 GWh
Supplied (gross): 108 771 GWh

Hence the average efficiency of generation is 108 771/228 242 = 47.7%.

In addition, allowance needs to be made for grid electricity losses. These will vary with location in the country and the level of voltage supplied.

Table 5.1 of DUKES gives:
Final consumption: 285 835 GWh
Losses: 26 305 GWh

So as a percentage of the electricity generated the UK-wide losses are 8.4%. At times of high demand the losses will be higher than this as part of the loss is proportional to the square of the current (I^2R). The loss at the point of connection of a CHP plant or heat pump supplying a large DH network will be a proportion of the total as part of the losses occur in the LV distribution system and a large DH energy centre is normally connected at HV.

For the purposes of the analysis below a loss figure of 5% has been assumed as representative of the loss to the point of connection of CHP or heat pumps for DH purposes.

Applying the 5% figure results in an average efficiency for grid supplied electricity of 0.95 × 47.7% = 45.3%. Grid efficiency is given the symbol η_{grid} below.

Gas-fired CHP

For a CHP plant, the primary energy efficiency is defined as for boilers as:

$$\text{CHP PE efficiency} = \text{heat out/fuel in}$$

But this doesn't take account of the electricity generated. So the 'fuel in' needs to be the net fuel used after the fuel displaced by the electricity generated by the CHP has been subtracted.

Suppose a CHP plant has a fuel input F_{chp}, heat output H and electricity output E. Then,

$$\text{CHP PE efficiency} = H/\left(F_{\text{chp}} - F_{\text{grid}}\right)$$

$$\text{Grid efficiency} = \eta_{\text{grid}} = E/F_{\text{grid}} \text{ or } F_{\text{grid}} = E/\eta_{\text{grid}}$$

$$\text{So CHP PE efficiency} = H/(F_{\text{chp}} - E/\eta_{\text{grid}})$$

$$= (H/F_{\text{chp}})/(1 - E/(F_{\text{chp}}.\ \eta_{\text{grid}}))$$

CHP efficiencies are expressed as:

$$\eta_h \text{ for thermal efficiency} = H/F_{\text{chp}} \text{ and}$$

$$\eta_e \text{ for electrical efficiency} = E/F_{\text{chp}}$$

So

$$\text{CHP PE efficiency} = \eta_h/\left(1 - \eta_e/\eta_{\text{grid}}\right)$$

It can be seen that as the CHP electrical efficiency η_e drops to zero the CHP efficiency tends to the CHP thermal efficiency and if the CHP electrical efficiency was equal to the grid efficiency the CHP efficiency become infinitely high, in other words the heat is genuinely 'waste heat' with no fuel input required.

If we assume that there will always be losses of about 20% from a CHP plant, i.e., if $\eta_h + \eta_e = 80\%$ and assuming $\eta_{\text{grid}} = 45.3\%$ as calculated above then we can plot the CHP primary energy efficiency against its electrical efficiency as shown in figure 6.9.

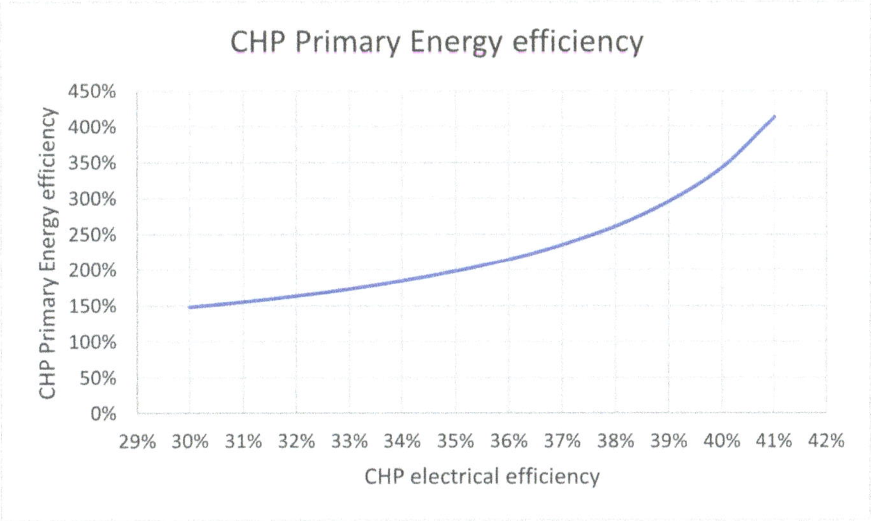

Figure 6.9. CHP Primary Energy Efficiency for heat production against CHP electrical efficiency.

Heat pump

For the heat pump we can define

Heat pump primary energy efficiency = heat out/fuel in

In this case the fuel in is the primary fuel used to produce the grid electricity.

A heat pump is defined by its coefficient of performance or CoP

$$\text{CoP} = \text{heat out/electricity in}$$

$$\text{and electricity in/fuel in} = \text{grid electricity efficiency, } \eta_{\text{grid}}$$

Hence:

$$\begin{aligned}\text{Heat pump PE efficiency} &= \text{heat out/fuel in}\\ &= \text{heat out/(electricity in}/\eta_{\text{grid}})\\ &= \text{CoP}.\ \eta_{\text{grid}}\end{aligned}$$

Assuming $\eta_{grid} = 45.3\%$ the heat pump efficiency against CoP is shown in figure 6.10.

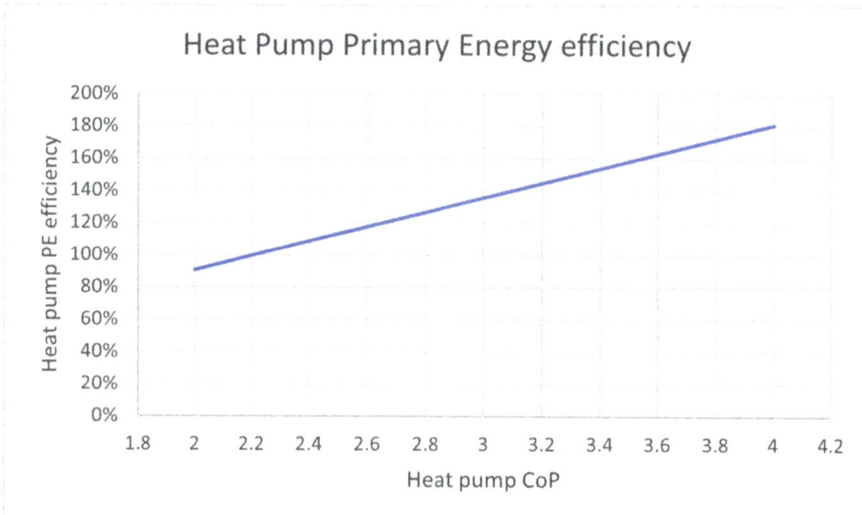

Figure 6.10. Heat pump primary energy efficiency against CoP.

Figures 6.9 and 6.10 can be used to assess primary energy efficiency and compare this with a boiler efficiency of say 80%. The data for a typical boiler, heat pump and CHP is given in figure 6.11.

It is also possible to compare the technologies using the inverse of efficiency which some may find easier to understand as it avoids having efficiencies greater than 100% which is counterintutive. For power plant this is known as the heat rate, i.e., fuel used per unit of electricity produced. To avoid confusion with heat rate, this metric is defined here as:

$$\text{Fuel use rate} = \text{primary fuel used} \times 100/\text{useful heat delivered } (\%)$$

The results for the fuel use rate for a typical boiler, heat pump and CHP are plotted on figure 6.12 for comparison purposes.

Figure 6.13 represents CHP and heat pumps on the same diagram and shows that in essence both technologies involve a change in the fuel used in the electricity system.

In the centre of this diagram the electricity and heat demand is shown. On the right-hand part of the diagram, the electricity demand is met with a fuel input F_0. The heat demand H is met by a heat pump and additional electricity is required, the volume of which is defined by the CoP for the heat pump. In the left-hand part of the diagram the same fuel input F_0 is used for the CHP plant. The heat demand is met by CHP which produces some electricity but not enough to meet the electricity demand

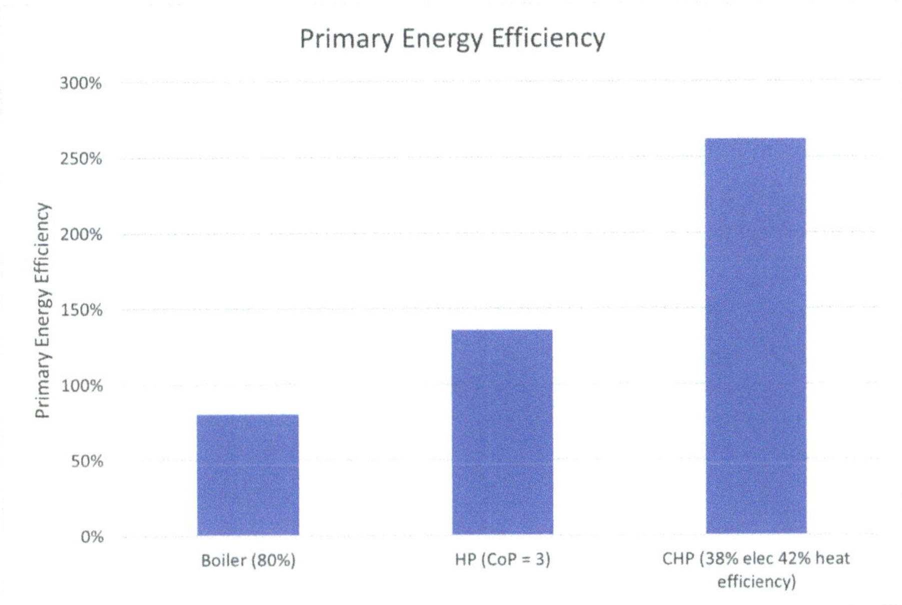

Figure 6.11. Primary energy efficiency of typical boiler, heat pump and CHP.

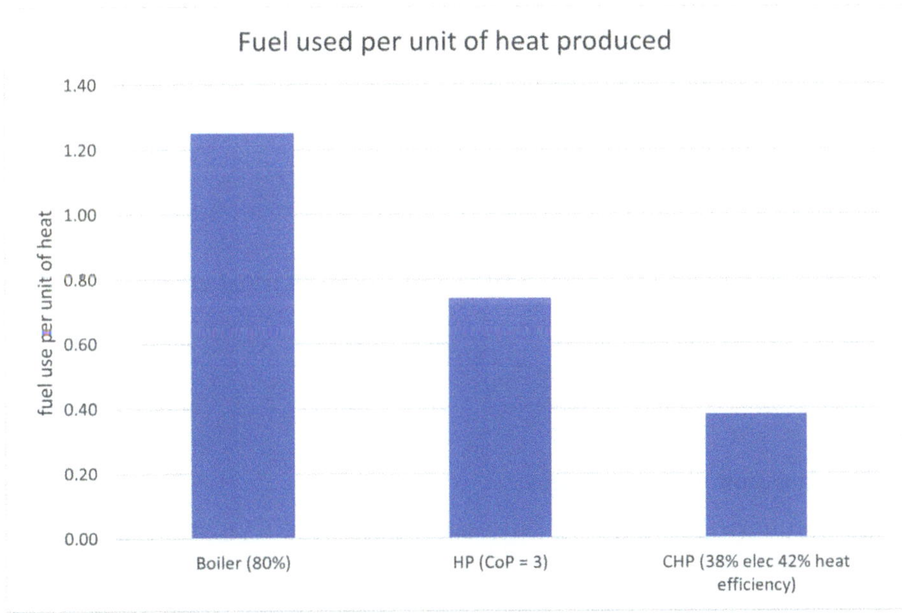

Figure 6.12. Fuel use rate of boilers, heat pump and CHP.

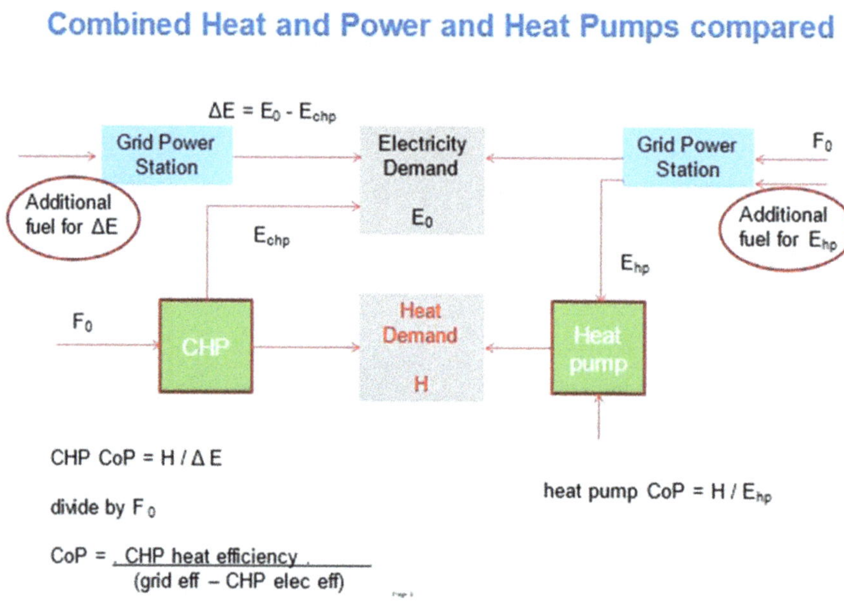

Figure 6.13. Comparison based on equivalent CoP of CHP.

as it is less efficient than grid power stations. So additional electricity, ΔE is needed to make up this difference.

It is therefore possible to define an equivalent CoP for a CHP plant as:

$$\mathrm{CoP_{chp}} = H/\Delta E = H/\left(E_0 - E_{\mathrm{chp}}\right)$$

Dividing through by F_0

$$\mathrm{CoP_{chp}} = (H/F_0)/\left(E_0/F_0 - E_{\mathrm{chp}}/F_0\right)$$

$$\mathrm{CoP_{chp}} = \eta_{\mathrm{h}}/(\eta_{\mathrm{grid}} - \eta_e)$$

For example, if the grid efficiency is 45%, the CHP electrical efficiency is 38% and heat efficiency 40%,

$$\mathrm{CoP_{chp}} = 42/(45 - 38) = 6$$

6.4.3 CO_2 emission per unit of heat

The second comparison is similar but uses the metric of CO_2 emission per kWh of heat supplied.

Grid electricity

Using the grid efficiency of 45.3% calculated in section 6.4.2, the CO_2 emission of electricity delivered can be calculated assuming a gas emission factor. For compliance with Building Regulations the CO_2 emissions need to be calculated and for this purpose the SAP model is used. The gas emission factor used in SAP is 0.21 kg kW^{-1} h^{-1} [12] which includes an allowance for upstream emissions. This is higher than the figure of 0.183 kg kW^{-1} h^{-1} which represents the CO_2 emissions associated with direct combustion (see section 9.2.1). In this section the SAP figure has been used.

Hence:

Gas emission factor = 0.21 kg kWh^{-1}

Electricity emission factor for CCGT = $0.21/\eta_{grid}$ = 0.21/0.453 = 0.46 kg kWh^{-1}

CHP plant

Suppose a gas-fired CHP plant has a fuel input *F*, heat output *H* and electricity output *E* and:

Electricity emission factor e_e

Gas emission factor e_g

CO_2 emission per unit of heat is the emission associated with the CHP fuel less the emission displaced by the electricity generated

$$CO_2 \text{ emission per unit of heat} = \left(F.\ e_g - E.\ e_e\right)/H$$

Dividing through *F* this can be expressed as:

$$CO_2 \text{ emission per unit of heat} = \left(e_g - E/F \times e_e\right)/H/F$$

But *E*/*F* and *H*/*F* are the CHP electrical and heat efficiencies, hence:

$$\mathbf{CO_2 \text{ emission per unit of heat} = (e_g - \eta_e \times e_e)/\eta_h}$$

Heat pump

For a heat pump the CO_2 emissions are simply the electricity emission factor divided by the CoP

$$\mathbf{CO_2 \text{ emission per unit of heat for heat pump} = e_e/CoP}$$

In the short-term when gas-fired CCGT power plant is contributing significantly to the grid electricity production then gas-fired CHP will have this CO_2 content of heat at any time when a CCGT is operating. This is because the CCGT plants are the marginal plant which will vary output in relation to demand. When a CHP plant operates and generates electricity the grid demand falls and it is the marginal CCGT plant that will reduce its output—the output of solar, wind and nuclear will be unaffected as their marginal operating costs are much lower than a CCGT plant and will continue running to maximise output.

Similarly, if a heat pump is operated, demand will increase and the output of CCGT plant will increase with no change in the output of solar, wind and nuclear.

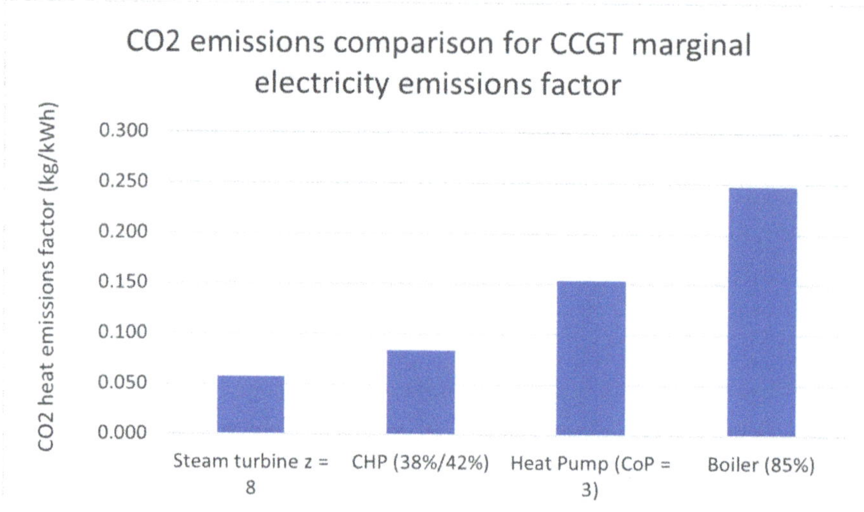

Figure 6.14. Comparison of CO_2 emissions assuming marginal electricity emissions factor of 0.46 kg kW^{-1} h^{-1} (assumes CCGT is the marginal plant).

In this scenario, a comparison between gas-engine CHP and heat pumps shows that CHP will deliver lower CO_2 emissions than the heat pump, see figure 6.14. The figure also shows the emission rate for heat extraction from a steam turbine which has the lowest emission rate (see section 6.5).

However, for a future scenario at times when there is potential surplus low carbon electricity on the system the CO_2 emissions from heat from CHP will be about double that from boilers, as there will be no CO_2 displaced at grid power stations. For DH schemes using CHP it will be important to make sure the plant can operate flexibly so as to avoid operating at times when CCGT plant is not operating and renewables are the marginal plant. This is likely to occur first in periods of high winds, high solar output and low demand.

For DH schemes using heat pumps the aim will be to operate at times when there is a potential surplus of renewable energy, which may occur more often at night and so thermal storage will be important and this may need to be increased above that typically used with CHP plant. The operation of CHP and heat pumps and the use of thermal storage is discussed further in section 10.2.

6.5 High temperature heat recovery from power plant and process plant

This section discusses technologies where heat is recovered at a high enough temperature to be used directly in a DH network. Although some of these examples could be seen as CHP plant it is clear that the primary purpose of the plant is for another purpose, e.g., to dispose of waste or generate electricity and the heat is seen as a byproduct. This is an important distinction in terms of the decision making process involved and the way in which the heat is recovered.

Combined cycle gas turbine

For larger DH projects combined cycle gas turbine (CCGT) plant can be used to supply heat to DH as part of an integrated national energy strategy. Normally the CCGT will be operated to generate throughout the year independent of heat demand and so will provide most of the heat from extraction of steam from the steam turbine. The steam will typically be extracted at around 1 bar a to provide temperature in the DH circuit of ca. 90 °C. Steam that is not needed for heating will continue to expand through the lower pressure stages of the steam turbine before being condensed and returned to the boiler.

Extraction of steam in this way releases a large amount of heat but at the expense of electricity generation which will be reduced as there is less steam flowing through the lower stages of the turbine.

The ratio between heat extracted and electricity reduction is approximately constant but will vary with the pressure at which the steam is extracted, the temperature at which the steam would otherwise be condensed and the amount of steam extracted. The ratio is known as the *z*-factor:

$$z = \text{heat extracted/reduction in electricity output}$$

For turbines designed for DH, the *z*-factor is typically in the range of 6–10. Research to support the development of the SAP 10.2 model has established a default *z*-factor figure of 10 [13]. It can be seen that this ratio is equivalent to the CoP ratio for heat pumps which is typically about 3, so extraction from a steam turbine is a very efficient process. The CO_2 emissions for this heat are the emissions associated with the reduction in electricity that needs to be generated on the grid to compensate, hence:

$$\mathbf{CO_2\ emission\ per\ kWh\ for\ the\ heat\ extracted} = e_{\mathbf{grid}}/\mathbf{z}$$

(see also figure 6.13 in section 6.4.3).

A second smaller source of heat from the CCGT is the exhaust gas. Depending on the design of the CCGT the temperature of the exhaust gases may be high enough to extract additional heat through what is known as a DH economiser. This will reduce the temperature of the exhaust gases below that which could be achieved with the steam raising plant. This additional heat recovery does not have a major impact on electricity generation although it does increase the resistance in the flue gas circuit and so results in a marginal increase in fan power.

CCGT plant are expected to be phased out before 2050 but they could be a future heat source for DH if carbon capture and storage technology is developed (see section 10.4).

Typical examples of CCGT supplying DH networks are:

Helsinki, Finland, Vuossari CCGT, 665 MWe, 582 MW thermal (https://www.helen.fi/en/about-us/energy/energy-production/power-plants/vuosaari-power-plants)

Malmo, Sweden, Oresundsvaeket CCGT, 440 MWe, 250 MW thermal (https://www.power-technology.com/projects/oresundsverket/)

Energy from waste

Energy from waste (EfW) plants use residual waste streams to generate heat or electricity or both. Early DH schemes often used waste as a fuel just to generate heat (this is how Sheffield Heat and Power began operating) but it is normally more cost effective to use the waste as a fuel in a boiler to raise steam to drive a turbine which generates electricity.

Heat is recovered in a similar way to CCGT mainly through steam turbine extraction. The SAP methodology uses a z-factor of 10 as a default value for EfW plants, based on performance taken from an existing EfW plant [13].

In some cases, additional heat is recovered from flue gases, especially where the heat from the waste plant is providing the base load for the DH. As the waste fuel typically has a high water quantity, the water vapour in the flue gas is significant and recovery of the latent heat in this part of the flue gas often justifies the additional investment in flue gas condensation heat exchange equipment.

The main drawback for these plants is that air quality concerns mean that the waste plant may be sited some distance outside the town or city and so a transmission main will be needed to deliver heat into the city area.

There are now a number of UK examples of energy from waste plants supplying DH networks as well as recently constructed continental projects:

Southwark, London, SELCHP, South East London Combined Heat and Power (https://selchp.com/energy-recovery/district-heating/)

Leeds, Recycling and Energy Recovery facility Recycling and energy recovery facility (https://www.leeds.gov.uk/residents/bins-and-recycling/your-bins/future-of-leeds-waste)

Vantaa, Finland, (https://www.vantaanenergia.fi/en/waste-to-energy-gives-a-new-life-for-rubbish/)

Copenhagen, Denmark Amager Bakke (Copenhill) (https://a-r-c.dk/amager-bakke/)

Biomass CHP

Biomass can also be used as a fuel to generate heat or electricity or both. Larger schemes will be profitable to generate electricity as well as heat (CHP) especially where financial incentives are available for renewable electricity generation.

Boilers may have grate firing or utilise fluidised bed firing.

For conventional combustion and steam turbine generation plants, heat recovery is similar to CCGT and energy from waste plants using extraction of low pressure steam from the steam turbine.

A UK example is:

Sheffield, Blackburn Meadows biomass CHP plant, (https://www.eonenergy.com/business/why-eon/case-studies/blackburn-meadows.html)

Advanced waste and biomass plants

Alternative technologies to conventional combustion, steam raising boilers and steam turbines have been supported by government incentives recently, e.g., pyrolysis and gasification. The benefits are:

- Higher electrical efficiency
- Lower emissions to atmosphere

- Useful by products
- Carbon capture in the form of biochar

Both of these processes will result in waste heat being rejected from the plant which can be used for DH. The technologies are at an early stage of development and may suffer from low availability which would be of concern to the DH operator.

Nuclear CHP

Nuclear power stations utilise a steam cycle to generate electricity and a steam extraction system to supply DH could be incorporated. The main drawback is that nuclear plants tend to be located some distance from population centres for safety reasons so a long transmission main would be required. This can only be justified if the demand is sufficiently large. So this source could be a long-term solution once the demand has been built up using smaller scale systems. Of the existing sites, Hinckley Point to supply Bristol, and Bradwell to supply London would be technically feasible. It is possible that the small-scale modular reactors currently being evaluated could also be used for DH and being smaller these may be located closer to cities.

An example of nuclear CHP supplying district heating is in Switzerland:

Beznau Switzerland, https://www.swissnuclear.ch/en/beznau-1-and-2-npp-_content—1–1036.html

Industrial waste heat

The above sources are all related to power plant operating in CHP mode. Some DH systems obtain high temperature heat from industrial heat recovery. These include oil refineries, cement works, steel works.

The proximity of industry to heat loads is a potential issue. In addition, on some sites the heat is rejected in a number of locations across the industrial site so a heat recovery network needs to be constructed through the site itself which can be difficult to deliver.

For this source of heat lowering the return temperature from the DH is particularly important. For a waste heat source in the form of water at 90 °C for example, the amount of heat that can be obtained is directly related to the DH return temperature. So with a DH return temperature of 55 °C the waste stream could be cooled to 60 °C but with a DH return temperature of 40 °C the waste stream could be cooled to 45 °C. This would mean 50% more heat available which may make the difference in commercial viability.

Commercially it would be risky to base a DH network solely on supply of heat from one industrial site as the industry and its processes may change over time. However, a number of DH schemes use industrial heat recovery as an important part of their base load heat supply.

Examples of industrial heat recovery supplying DH are:

Goteborg, Sweden using heat from oil refinery (https://smartcitysweden.com/best-practice/309/gothenburg-takes-district-heating-to-a-new-level-of-sustainability/#:~:text=Some%2070%20percent%20of%20Gothenburg%E2%80%99s%20district%20heating%20comprises,is%20fossil%20energy.%20Going%20fossil-free%20by%20storing%20heat)

Aalborg, Denmark using heat from cement works (https://www.cementirholding.com/en/sustainability/district-heating-and-cooling-in-aalborg-denmark)

Box 6.3. Cardiff district heating.

Project Name: Cardiff Heat Network

Local Authority: Cardiff City Council

Developer: Cardiff City Council

Construction Contractor: Pinnacle

Operator: Pinnacle

Lead consultants: Carbon Trust and Sustainable Energy Ltd

Energy from waste plant and above ground heat mains (Reproduced with permission from Sustainable Energy Ltd.)

Key data

Peak heat demand (MW):	15
Annual heat delivered to customers (MWh):	11 000 first year to 50 000 in later phases
Heat sources:	Steam extraction from energy-from-waste plant, gas-fired boilers
Date construction commenced:	2022
Date first heat delivered:	2024

Description

The Cardiff district heating scheme was developed by Cardiff City Council and is expected to start operating in 2024. The initial customers of the scheme include major buildings in the Cardiff Bay area including the Senedd (Welsh Government), the Wales Millennium Centre and the Council offices.

Heat is taken from the pre-existing energy-from-waste plant owned and operated by Viridor. This is a 30 MW plant that takes residual waste from the South Wales area. Standby gas-fired boilers are installed nearby under a major road flyover. Within the plantroom there are also four 30 m^3 thermal stores.

The heat network consists mainly of flow and return pre-insulated steel pipes with some sections of the main DN300 pipework running above ground. The main challenge in the construction phase was crossing a major road interchange, coordinating with new developments and working within government land. The system operates with a peak flow temperature of 90 °C which is reduced to 75 °C in summer. Return temperatures are expected to be 60 °C in summer and 55 °C in winter.

There is potential for significant expansion to the north into the City Centre areas.

Information kindly supplied by Sustainable Energy Ltd

6.6 Low temperature heat recovery with heat pump

A heat source for DH may come from another process which rejects heat to the environment but at a temperature lower than required for DH but by using a heat pump the supply temperature can be increased to suit the DH. The heat source could be from an industrial process or a commercial process such as cooling of a data centre. The obvious disadvantage is that it will be location specific which may not suit the proximity of the heat customers. It will also be considered a relatively high risk source in commercial terms as it will depend on the third party remaining in place for a long period. Provided the heat is rejected at a higher temperature than ambient water or air, especially in winter, this will result in a more efficient heat pump plant. Use of heat rejected from chillers can be part of an Energy Centre supplying DH and DC. This option is discussed further in section 10.7.

Datacentres are increasing in number and scale to meet the demands of the internet and require significant cooling. This creates a potential concentrated source of heat, often in the form of warm water at 20 °C–30 °C which is suitable for a heat pump application. Other similar sites that may be suitable include refrigerated stores and food processing.

Heat in the form of hot air may be available from ventilation shafts as in the Bunhill Phase 2 scheme in Islington (https://www.islington.gov.uk/environment-and-energy/energy/bunhill-heat-network)

Ground water may be extracted from former mineworkings at an elevated temperature although frequently the water is highly corrosive and will either need treatment before discharging or will need to be returned back into the mine at a different point to avoid recirculation. In some cases, mine water is being pumped out already to prevent contamination of water supplies and in these cases there is no additional energy use to make use of the mine water as a heat source for DH. The Heerlen scheme in The Netherlands is an example of this concept (https://guidetodistrictheating.eu/heerlen/). In the UK, the Gateshead district energy network is using mine water as a heat source (https://tp-heatnetworks.org/gateshead-mine-water-heat-network-complete/) and a new-build housing project at Seaham in County Durham will be based on using water taken from an existing mine water pumping and treatment plant (https://www.gov.uk/government/news/uks-first-district-heating-scheme-using-mine-water-energy-now-in-development)

Power stations are also a source of low temperature heat at around 30 °C. It would be more efficient to extract steam from the steam turbine but where the power station is an existing plant the necessary modifications to the turbine may be too costly, especially if downtime has to be taken into account. The alternative is to use the steam turbine condenser circuit as a source of low temperature heat for a heat pump installation. This is only practical where the condenser is a water-cooled rather than an air-cooled condenser circuit. The disadvantage of this source is that in most cases the power station will be some distance away from heat loads and so a sufficiently large heat network will have to be built up using other sources first. The energy efficiency benefits may not be enough to outweigh the cost of the transmission main. Also, in the future fossil fueled power stations may operate relatively few hours in the year (unless fitted with carbon capture, see section 10.4).

Box 6.4. South Dublin District Heating—data centre heat pump.

Project Name: Tallaght District Heating-Heatworks

Local Authority: South Dublin County Council

Developer: South Dublin County Council

Construction Contractor and Operator: Fortum and Enerpower

Lead consultant: Codema

Reproduced with permission from Fortum. (The DH Energy Centre is bottom left, dwarfed by the data centre to the right)

Key data

Peak heat demand (MW):	3
Heat sources:	Heat pump using data centre heat rejection, electrode boilers
Date construction commenced:	May 2021
Date first heat delivered:	January 2023

Description

The Tallaght district heating scheme was developed by South Dublin County Council from 2020 onwards. The customers of the scheme include the County Hall complex, and the Technical University of Dublin's Tallaght campus.

The primary heat source is two 1500 kW heat pumps connected in series. The heat pumps use heat rejected from the adjacent large Amazon Web Services data centre as its source. Electrode boilers are installed for standby use. There is permission for two 200 m^3 thermal stores to be added in the future. The project will save 1500 tonnes CO_2 p.a.

The heat network consists of flow and return pre-insulated steel pipes with some sections in twin pipes. The main challenge in the construction phase was installing the heat mains in a busy part of Dublin. The system operates with a peak flow temperature of 80 °C which is reduced to 70 °C in summer. Return temperatures are currently around 50 °C.

Plans for the future include the extension of the network to supply a large housing development. Further information is available at https://heatworks.i.e./

Information kindly supplied by Codema

Waste water and sewage flows can be used as a source for heat pumps as they are typically at slightly elevated temperatures compared to ground or surface water systems. At sewage works treated water is discharged back to rivers and represents a concentrated heat source well suited to a DH installation.

6.7 Renewable heat sources

6.7.1 Biomass boilers

Biomass boilers are well suited to supply DH schemes, especially by providing the base-load heat. Costs of boilers fall with scale and it is usually easier to incorporate thermal stores and fuel stores at a central energy centre location than on high value land or in buildings. Concerns over air quality can be addressed more easily at larger centralised plant where tall stacks may be more acceptable, but even so they are unlikely to be acceptable within city boundaries. There are also concerns about additional HGV journeys for fuel deliveries.

The main concern over the use of biomass is whether the CO_2 emitted will be absorbed again by growth of the replacement biomass and in a short enough timescale to combat climate change. The change in land use also needs to be taken into account to ensure that the stock of trees is not being depleted.

The Government has set up a Timber Standard [14] which suppliers to projects supported by the Renewables Obligation and the Renewable Heat Incentive must comply with. This aims to ensure that all wood fuel is obtained from legal and sustainable sources.

As biomass is relatively bulky the energy used in transport should be considered. This implies using large lorries to minimise the number of deliveries and to utilise local sources wherever possible.

As biomass is a fuel that is capable of producing high temperatures and has limited availability, the most effective use for biomass is in CHP plant, preferably fitted with carbon capture rather than heat only boilers (see section 6.5).

UK examples include:

Queen Elizabeth Olympic Park where a 3 MWth biomass boiler firing on woodchips supplies the base load for the Park; the fuel must be sourced within a 100 km radius of the site (https://www.queenelizabetholympicpark.co.uk/about-us/environmental-sustainability/climate-emergency)

St Andrews University where a biomass boiler remote from St Andrews supplies the University via a 6 km pipeline; the fuel is woodchip derived from local forestry operations (https://www.sustainabilityexchange.ac.uk/files/disctrict_heating_-_a_technical_approach.pdf)

Further information on the design of biomass heating is given in CIBSE AM15 [15].

6.7.2 Deep geothermal

The use of deep geothermal boreholes to access heat from >1000 m below the surface is very suitable for DH as the scale and cost of the borehole is significant and so supplying multiple buildings will be needed if the project is to be economically viable. In the UK only certain areas are expected to yield sufficiently high temperatures, e.g., Cornwall and Cheshire.

At sufficient depth, ground water can be extracted at high enough temperature to be used directly to heat a DH circuit. It is not carbon free as energy is required to pump the water to the surface. Even if the temperature is not high enough for direct DH use, a heat pump can be installed to lift the temperature further.

The main disadvantage is risk associated with very high costs coupled with uncertainty involved in the drilling process as predictions of water volume flow rate and temperature will be liable to error.

As for industrial heat recovery, a low return temperature in the DH network will be of benefit to maximise the energy available for a given well. Given the majority of the project cost is in the drilling of the well, the size of the resource is critical and for a given water flow this is determined by the DH return temperature.

6.7.3 Solar thermal

The conversion of solar energy to heat is a very efficient process in terms of energy yield per m^2. It is about three times as efficient as conversion to electricity using photovoltaic cells and at least 30 times better than the conversion to heat via biomass (https://solarthermalworld.org/news/solar-thermal-shows-highest-energy-yield-square-metre/). The capital costs per kW for a DH scale solar heating plant are significantly lower than for an individual building solar water system.

Without interseasonal storage the heat would be available mainly in summer. For most schemes, there will be other primary low carbon plant sized to supply most of the winter demand which will have excess capacity in summer and so the solar thermal will have relatively little additional benefit in lowering costs or carbon and yet incur a high capital cost. For this reason, solar thermal is used in conjunction with an interseasonal heat store that allows the heat obtained in the summer to be used in winter.

An interseasonal store is usually in the form of a shallow pit using similar technology to that used for landfill of waste with a rubber liner to retain the water. The use of a below ground pit avoids the cost of constructing the walls and floor of a store. An insulated cover is required and this can consist of floating polystyrene blocks. Other methods such as aquifer thermal energy storage are also used and are discussed further in section 10.2.5.

The main disadvantage is the land area required for both the solar thermal panels and the interseasonal heat store.

Temperatures from the solar panels can exceed 70 °C which even after losses through the year will be sufficient for modern DH schemes. In some cases heat pumps may be needed to boost the temperatures further.

An example from Denmark is the Marstal scheme, which includes a 75 000 m^3 interseasonal store (https://www.euroheat.org/resource/marstal-district-heating.html).

A further option for solar energy is to extract heat from the power turbine associated with a concentrated solar power project. This is still an emerging technology.

Box 6.5. Queen Elizabeth Olympic Park.

Project Name: Queen Elizabeth Olympic Park (East London Energy)

Local Authority: London Legacy Development Corporation

Developer: Equans

Construction Contractor: Equans

Operator: Equans

This Energy Centre, Olympic Park - geograph.org.uk - 3118137.jpg image has been obtained by the author(s) from the Wikimedia website where it was made available under a CC BY-SA 2.0 licence. It is included within this book on that basis. It is attributed to Energy Centre, Olympic Park by Oast House Archive.

Key data

Peak heat demand (MW):	36
Annual heat delivered (MWh):	109 000 MWh
Annual cooling delivered (MWh):	49 000 MWh

Heat sources:	• Energy Centre 1 (Kings Yard)
	  • 3.1 MWe CHP
	  • 3.5 MW biomass boiler
	  • 40 MW conventional boilers
	  • 4.0 MW absorption chiller
	  • 14 MW vapour compression chillers
	• Energy Centre 2 (Stratford City)
	  • 2 ×3.1 MWe CHP
	  • 40 MW conventional boilers
	  • 4.0 MW absorption chiller
	  • 35 MW vapour compression chillers
	  • 3.2 MW water source heat pump

Date construction commenced:	2008
Date first heat delivered:	2010

Description

The concept of the ELE district energy system was developed by the Olympic Delivery Authority (ODA), now the London Legacy Development Corporation (LLDC), and Stratford City Development Limited (SCDL) to supply low cost, low carbon heating and chilled water to buildings across the Queen Elizabeth Olympic Park (QEOP) and surrounding areas. Equans (formerly Cofely) was awarded a 40-year contract by the ODA to develop the system, with Equans providing funding to design and build the scheme and then provide operation, maintenance and associated metering and billing services for the term. As part of the form of agreement with the LLDC, new developments 'on-park' are obligated to connect to the district energy system under pre-agreed commercial terms, including a regulated tariff and indexation structure.

The scheme currently operates with base load heat provided by the CHPs and biomass boiler, with back-up and top-up provided by the gas-fired boilers. The cooling is generated from a combination of absorption chillers, supplied with heat from the CHPs, and electric vapour compression chillers, supplied with electricity from CHP generation and grid import. There are also 3 × 750 m^3 thermal stores.

The scheme includes over 20 km of heat network with over 5500 residential customers. ELE provides a source of more than 75 MW of low carbon heating and hot water, 30 MW of chilled water and 6 MWe of power across the Queen Elizabeth Olympic Park and the surrounding area; playing a fundamental role in continuing the London 2012 legacy strategy, the scheme serves an area of over 2.5 km^2. The system operates with a peak flow temperature of 90 °C and 55 °C return temperature.

The district heating network serves both 'on-park' and 'off-park' customers. ELE is currently serving over 12 GWh of heat from biomass alone. Phase 1 saw EQUANS supply energy to the sporting venues and athletes' village (later to become East Village) for the London 2012 Olympic and Paralympic Games and the new Westfield Shopping Centre. Phase 2 has seen the addition of connections to the existing network to serve the new businesses and neighbourhoods that have been developed post-games on the park and in the surrounding areas.

The scheme also includes a district cooling system, with over 2 km of network, mainly from the Stratford energy centre and supplying the surrounding Westfield buildings.

Energy Centre Construction began in October 2008 which included two new purpose-built energy centres: Kings Yard for Olympics and Stratford for Westfield Development. All major plant items were installed by March 2010.

As part of the decarbonisation strategy on ELE, Equans have developed a Roadmap to transition the scheme away from fossil fuels. In collaboration with the LLDC and SCDL, Equans have developed a series of projects which will achieve site wide decarbonisation via a mix of system optimisation and incremental decarbonisation activities to reduce temperatures on the network and introduce heat pumps. Equans have committed to all of their city centre schemes to be net zero ready by 2035. The first step in the transition to net zero at ELE is underway with a 3.2 MW heating/2 MW cooling water source heat pump, which will utilise heat rejected from the chillers as its heat source, to be installed at Stratford City Energy Centre and operational in 2023. As a result, the gas usage from boilers and then CHPs is offset, with the heat pump contributing about 18% of the annual heat supply and resulting in a 15% reduction in total carbon emissions.

Information kindly supplied by Equans

6.8 Summary of low carbon heat production technologies

A DH system which can be implemented in any location is currently likely to use either gas-fired CHP or centralised air source heat pumps or a combination of both, together with thermal storage.

A gas-engine CHP system can have a lower operating cost because income is obtained from the electricity generated to offset the fuel and the capital costs. In contrast, heat from a heat pump system will have a higher cost as it needs to cover both the capital cost and the cost of electricity for the heat pump.

As the grid decarbonises, the operating hours for gas-fired CHP will need to be progressively reduced to match the times when CCGT is operating on the grid as otherwise at other times CO_2 emissions from CHP would be double that from boilers. This reduction in operating hours needs to be taken into account when assessing the case for new CHP plant (see also section 6.9).

As outlined in this chapter there are a number of other heat sources that are site specific but that could offer advantages over the gas CHP and ASHP base case.

All of the low carbon heat sources discussed in chapter 6 are summarised in table 6.2 below with qualitative commentary where possible in relation to costs and environmental impacts. A typical feasibility study will review all of these options and their suitability for the planned DH scheme. The heat sources are grouped as follows:

- Gas-fired local CHP
- Heat pumps using ambient sources
- High grade heat recovery
- Low grade heat recovery with heat pumps
- Renewable heat sources

Table 6.2. Summary of low carbon heat sources used in DH.

Heat source	Capex impact	Opex impact	CO_{2e} impact	Comments and examples
Combined heat and power				
Gas-engine CHP	Higher than ASHP for same heat capacity	Benefits from sales of electricity, especially if private wire	Good but only of benefit when fossil fuel power stations are displaced on grid so may be phased out by 2035	Many examples: Queen Elizabeth Olympic Park, Kings Cross, Gateshead and on campus sites, e.g., universities and hospitals
Fuel cell CHP	Higher than gas-engine	May be more efficient than gas-engine but operating constraints and lifetime limitations	Slightly better than gas CHP; only of benefit when fossil fuel power stations are displaced on grid so may be phased out by 2035	Quadrant 3, Regent St
Heat pumps using ambient heat sources				
Air source heat pump	Lowest cost heat pump solution	CoP depends on refrigerant selection and may be poor in very cold weather	Good and will benefit further at times when renewable electricity is in surplus	Chilton Woods housing scheme, Solihull Town Centre (in development)
River/lake/canal/harbour/sea water source heat pump	Typically higher due to cost of extraction equipment, filtration, less space required than ASHP	Slightly lower than ASHP as higher source temperature in winter (but lower in summer) and additional pumping energy needed	Some additional benefit cf ASHP especially in winter	Many cities developed as ports so river water may be available however practicality of extraction of water may limit benefit. Examples: London, Bristol and Glasgow
Ground water source heat pump—borehole	Higher due to cost of drilling borehole	Slightly lower than ASHP as higher source temperature in winter (but lower in summer) and additional pumping energy needed	Some additional benefit cf ASHP especially in winter	Risks include insufficient flow rate, reduction of flow over time, uncertainty of drilling costs

Ground water source heat pump—closed loop	Higher due to cost of drilling boreholes, large ground area required	Slightly lower than ASHP as higher source temperature in winter (but lower in summer)	Some additional benefit cf ASHP especially in winter	Risk of drilling costs, long-term reduction in ground temperature leading to loss of efficiency benefit. Need a large land area so more suitable for smaller schemes. May be optimised by combining with ASHP, solar thermal and interseasonal storage
High grade heat recovery				
Steam extraction from power station	Can be low if designed for DH but likely to need transmission main (which is feasible if heat demand is large enough)	If optimised for DH supply, very low as *z*-factor about 8	If designed for DH supply, can be very low as more than twice as efficient as ASHP	Existing steam turbines not designed for efficient CHP. Fossil fuelled power stations may have limited life due to decarbonisation unless using carbon capture. Nuclear stations likely to be sited away from cities
Steam extraction from energy from waste plant	Can be low if designed for DH but likely to need transmission main (which is feasible if heat demand is large enough)	If optimised for DH supply, very low as z-factor about 8	If designed for DH supply, can be very low as more than twice as efficient as ASHP	Many UK examples of successful schemes. But new EfW plants would need to be sited near to cities which raises air quality concerns and need to be designed from the outset to provide heat. Examples: Nottingham, Sheffield, SELCHP, Leeds and Cardiff
Industrial heat recovery—direct	Low as only need heat exchanger. However, may be high cost to collect heat from a number of discharge	Low as minimal electricity input but depends on negotiation with the site	Almost zero emissions as minimal electricity input	Commercially risky to base a DH investment on a single industrial source unless a very secure established business. Can be

(*Continued*)

Table 6.2. (*Continued*)

Heat source	Capex impact	Opex impact	CO_{2e} impact	Comments and examples
	points on the site and high heat exchanger cost if source is hot air. Also, may need a transmission main if at fringes of city			difficult to negotiate with industrial owner as perceived risk to main process activity. Examples: Aalborg (cement works) and Goteborg (oil refinery)
Low grade heat recovery				
Data centre heat rejection heat pump	May be lower than ASHP if no need for back-up heat source and supply of heat is secure enough, less space required than ASHP. Needs a site near to data centre	Significantly lower than ASHP as source temperature higher but depends on negotiation with the site	Significant additional benefit cf ASHP	Data centre owner may charge for the heat. Datacentres often located in cities where electricity distribution system is strong. Example: South Dublin Tallaght scheme
Industrial heat recovery—with heat pump	May be lower than ASHP if no need for back-up heat source and supply of heat is secure enough, less space required than ASHP. Also may need a transmission main if at fringes of city	Significantly lower than ASHP as source temperature higher but depends on negotiation with the site	Significant additional benefit cf ASHP but depends on temperatures available	Risky to base a DH investment on a single industrial source unless a very secure established business. Can be difficult to negotiate with industrial owner as perceived risk to main process activity.
Mine water source heat pump	May be lower than ASHP if mine water is already being pumped to surface to protect water supplies, otherwise similar to ground water source borehole	Lower than ASHP as source temperature higher but depends on negotiation with the site	Small additional benefit cf ASHP depending on site conditions and depth of mine	Needs evidence of long-term need for pumping. Example: Heerlen Netherlands and Seaham UK (planned). Mine water often contaminated so needs treatment unless discharged back into mine

Power station heat recovery—with heat pump	Heat from condenser circuit ca. 30 °C so need a heat pump. May be lower than ASHP but likely to need a transmission main	Lower than ASHP as source temperature higher. Potential for low electricity price if based on power station export price	Significant additional benefit cf ASHP	Many modern CCGT/EfW plants have air-cooled condensers which makes heat recovery impractical. Fossil fueled power stations may have limited life due to decarbonisation unless using carbon capture. Nuclear stations likely to be sited away from cities
Waste water source heat pump	May be lower than ASHP if no need for back-up heat pump heat source and supply of heat is secure enough, less space required than ASHP, but likely to need a transmission main	Slightly lower than ASHP as source temperature higher in winter	Small additional benefit cf ASHP	Sewage treatment works likely to be on edge of city but are likely to have a long life and so heat source is secure. Water can be used directly as already treated for discharge into rivers. Discharging at lower temperature may have environmental benefit. Space may be available at treatment works
Renewable sources				
Biomass boilers	High for biomass boilers and fuel storage	Depends on biomass fuel price which is typically linked to fossil fuels	Depends on biomass source for fuel and	Concern over sustainability of fuel and land use impacts so actual CO_2 benefit uncertain. Examples:

(Continued)

Table 6.2. (*Continued*)

Heat source	Capex impact	Opex impact	CO_{2e} impact	Comments and examples
			timescale for replacing source	Queen Elizabeth Olympic Park and St Andrews University
Deep geothermal	Very high for costs of drilling to >1000 m	Very low assuming temperatures high enough not to need heat pump	Low as renewable energy source and limited electricity input	Uncertainty over temperatures obtained. Examples: Southampton, Cornwall hot dry rocks programme
Solar thermal and interseasonal storage	High cost for solar thermal panels and interseasonal store. Requires large land area. Will need a transmission main into the city	Low as heat pump may not be needed and minimal electricity input	Almost zero as renewable energy source	Numerous examples, e.g., in Vojens and Marstal, Denmark. Most suitable for low temperature networks. May be combined with closed or open boreholes and water source heat pumps

6.9 The value of multiple heat sources

One of the benefits of DH systems is that more complex plant can be installed which is not normally practical at a building level. Central plant costs are also lower for DH systems due to economies of scale. As a result, it can be economic to install a combination of:

- Heat pumps
- CHP
- Gas-fired peaking boilers
- Electrode peaking boilers
- Thermal storage

This provides significant operational flexibility compared to, e.g., an individual building system using heat pumps alone.

As an illustration of the potential benefits of using multiple heat sources, the comparison below considers the use of various combinations of the plant listed above in terms of CO_2 emissions as the proportion of time that renewables are the marginal plant on the grid. The assumptions are:

- Low carbon plant is sized to supply 90% of the heat over the year.
- Peak boilers supply 10% of heat over the year.
- The marginal electricity plant at times of peak boiler use is CCGT with an emissions factor of 0.46 kg kW^{-1} h^{-1} (peak boiler use likely to coincide with high electricity demand and often low wind output due to anticyclonic conditions and low solar output as in winter).
- Gas emissions factor taken as 0.21 kg kW^{-1} h^{-1}, boiler efficiency 85%.
- CHP performance and heat pump performance as in section 6.4
- The proportion of renewables at the margin is assumed to be spread uniformly through the year.

Five strategies have been compared in figure 6.15:

A—CHP plant and gas-fired peak boilers
B—Heat pump and electrode boilers
C—CHP and electrode boilers
D—Heat pump and gas boilers
E—CHP, heat pump and gas boilers

Strategy A—CHP and gas-fired boilers

Section 6.4 compared CHP and heat pumps for the case where gas-fired CCGT is being displaced. As the grid decarbonises, the times when renewable sources will be at the margin increases and at these times the CHP is assumed to be shutdown and the boilers used instead.

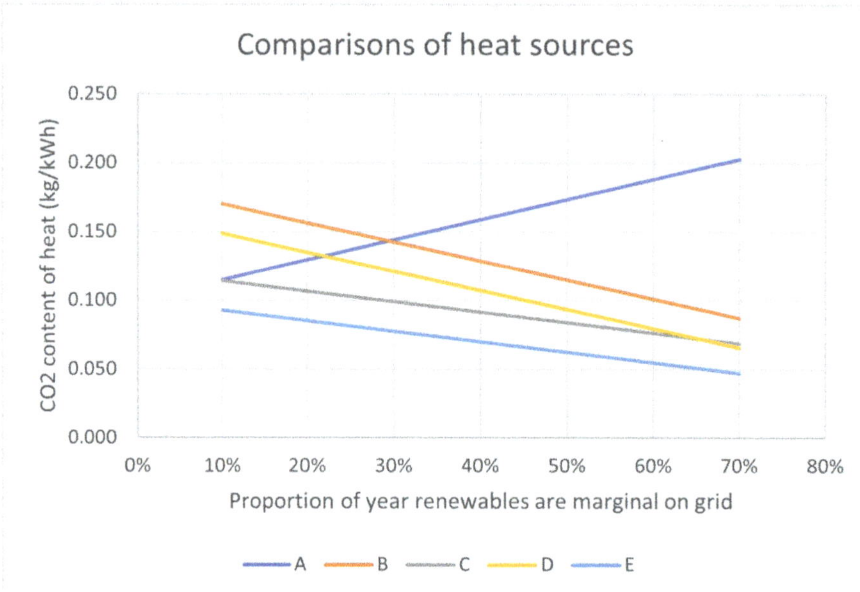

Figure 6.15. Comparisons of heat source strategies in terms of CO_2 content of heat.

As the renewables proportion increases the benefit available for CHP gets less and the CO_2 content rises.

Strategy B—heat pump and electrode boilers
This strategy has higher CO_2 content than the Strategy A CHP option but reduces as the renewables proportion increases and is lower than Strategy A for renewables proportion greater than 30%.

Strategy C—CHP and electrode boilers
This strategy maintains the benefit of CHP but when the CHP is shut down electrode boilers utilise the zero CO_2 marginal electricity and so the CO_2 content falls over time. This more than compensates for the high emissions for the 10% peak boiler use. The disadvantage of this strategy in practice is that the electrode boilers will be expensive to run.

Strategy D—heat pump and gas boilers
This strategy is similar to Strategy B but using gas boilers for the peak duty results in lower CO_2 than electrode boilers as CCGT is assumed to be the marginal plant at times of peak.

Strategy E—CHP, heat pump and gas boilers
For this strategy both CHP and heat pumps are installed and operated in a complementary manner. The CHP plant is used when the CCGT is the marginal plant and the heat pump is used when renewables are marginal. As the grid decarbonises the heat pump is used more and the CHP less. This strategy results

in the lowest CO_2 emissions for all proportions of renewables. The disadvantage is that the capital cost will be the highest if both low carbon sources are sized to operate to supply the 50% of peak demand. However, the operating costs are likely to be low as the use of CHP and heat pumps together with thermal storage will be able to fully exploit the time varying prices in the electricity market.

Further discussion on how the variability of renewable generation will impact the design of DH/C is given in section 10.2.

6.10 Optimal scale of DH networks

Research carried out by the IEA [16] showed that for gas-fired CHP, city-wide district heating supplied from large-scale CCGT plant would be the most economic and environmentally beneficial scale of development.

With heat pump systems, the scale of a DH network will be determined by the heat source for the heat pump. Some sources may be localised and of finite output, e.g., a river, or data centre, others may be uniformly dispersed such as ambient air.

With ASHP based systems, smaller networks are likely to prevail as there will be limited gains from economies of scale by using multiple heat pumps in one location. Eventually the smaller heat networks may be linked to improve security of supply so that a large DH network may have multiple heat sources supplying it.

Transmission mains

Some of the heat sources most suitable for DH are likely to be found on the fringes of cities and there will be a need for a transmission main to transport the heat into the higher density areas of the city. Long distance transmission mains are common on large continental schemes, often used to connect cities with large power stations. In common with other forms of piped transport for oil, gas and water this can be an economic method for transfer of heat but the economics depend very much on scale.

The cost of the transmission main needs to be recovered by the volume of heat sold. The greater the amount of heat transported the more viable the transmission main. This is because the capacity of the pipeline is proportional to the square of the diameter whereas costs do not increase at the same rate (see table 6.3). The capital cost for transporting 45 MW is less than half that of transporting 11 MW. Or to put

Table 6.3. Transmission mains at 200 and 400 mm diameter.

	200 mm diameter	400 mm diameter
Temperature difference (°C)	30	30
Velocity (m s^{-1})	2.5	3
Heat transport capacity (MW)	10.6	44.7
Cost per m (£/m)	1264	2270
Cost per m per MW of capacity (£/m/MW)	119	51

it another way, doubling the diameter allows four times the amount of heat to be transported for a little less than twice the cost.

This means that a transmission main will become more economic as the scale of the heat demand connected increases. Consequently, as the heat demand increases, heat sources from further away can be utilised economically. If zoning of city areas is introduced (see section 10.9), this will create greater certainty on the scale of the heat demand that can be supplied and hence widen the opportunities to use heat sources from further away that could provide larger volumes of lower carbon heat.

6.11 Peak and standby plant

6.11.1 Peaking plant

One of the benefits of DH systems is that a more complex heat production plant can be utilised that would be considered for an individual building system. One example of this is the use of peak duty plant.

A typical load duration curve for a DH system has already been presented in figure 2.10 together with the cumulative version of this in figure 2.11. This shows the proportion of annual heat consumption that occurs above a given heat demand.

For the example given, only 20% of the annual heat consumption occurs for demands above 30% of the peak and only 10% of the demand occurs for demands over 50% of the peak. Or expressed another way, 90% of the annual heat demand can be met by a primary production unit sized at 50% of the peak demand.

The benefit of this is that the capacity of the primary low carbon heat production plant can be substantially reduced rather than sizing this for the peak demand. The secondary plant which a lower capital cost per kW can be used to meet the peak demand.

For example, a CHP plant, heat pump or biomass boiler could be sized at 50% and a lower capital cost plant such as a gas boiler or electrode boiler could be sized for the remaining 50%.

Even if the operating costs of the secondary plant are much higher than the primary plant this will have a small impact on the overall economic case if the amount the secondary plant is used over the year is only 10%.

An economic optimisation study will seek to establish the best balance between the primary plant and the secondary peaking plant which will depend on the relative capital and operating costs of the two sources and the shape of the load duration curve. A value can be ascribed to the carbon saved to take into account the additional carbon savings as the primary plant capacity is increased.

At present gas-fired boilers are commonly used for peak duty as these are low cost and lower carbon than using electrode boilers as marginal electricity at times of peak heat demand is likely to be from gas-fired CCGT with CO_2 emissions about twice that of gas boilers. In the future when the grid has decarbonised there may be a case for electrode boilers as peak capacity. However, peak heat demands may coincide with periods of low wind output when it is more likely that CCGT plant will be in operation.

6.11.2 Standby plant

In addition to the use of primary production unit(s) and peak duty plant there may be a need for additional heat production units to provide standby supplies should one or more of the heat production units fail.

Whereas with an individual building such provision may seem excessive, with DH when a larger number of customers could be affected at the same time with significant potential for compensation claims it is normal to provide some standby.

The standby capacity to be provided is a complex question depending on:

- The probability of failure of a primary production unit
- The number of primary production units
- The capacity of the primary production units
- The probability of failure of the secondary plant
- The number of secondary production units
- The shape of the load curve and the required level of reliability for different demand levels

6.11.3 Location of peak and standby boilers in a network

It is often the case that peak and standby boilers are located at the same energy centre as the low carbon heat production plant. However, there are cases where the boilers could be distributed within the network. For example:

- Where the low carbon plant is some distance from the heat demand and a transmission main is required. It would be better to size the transmission main just for the maximum capacity of the low carbon plant and locate the peak boilers nearer the demand. This reduces the cost of the transmission main and improves security of supply.
- Where the customer has particular concerns over security of heat supply and the on-site boiler plant is going to be retained then these boilers could be used for peak duty as well. This will mean that the local network need not be sized to meet the peak demand of these buildings thus saving in capex.
- Where there is a large network, security of supply would be improved if the boilers were distributed within the network so that there would also be fewer single points of failure for the network and also parts of the network could be shutdown for maintenance without loss of heat supply to the entire network.

Where boilers are not located at the energy centre it will normally be necessary to install a small boiler at the energy centre to ensure that the network is still kept up to normal operating temperature at times when the low carbon plant is not in service. Otherwise, the cooling and heating cycles that might result would be damaging to the pipework (see section 5.7).

For more information on boilers and heating systems in general see CIBSE Guide B1 Heating [17].

6.12 Thermal storage

Thermal storage is well suited to being included in a DH scheme because the costs reduce and efficiency improves with scale.

Smaller UK schemes have used stores designed to work on a daily cycle to smooth out demand peaks and hence to minimise the capacity and use of peak boilers. Other advantages are:

- To enable operation of a CHP plant to take place at times of maximum electricity price.
- To allow a CHP plant to operate at maximum output rather than modulating to improve efficiency and minimise maintenance costs.
- To operate a heat pump at times of minimum electricity price or highest source temperature.
- To operate a group of heat pumps at optimal output rather than following the heat demand pattern.
- To maximise heat recovery from an industrial process to meet a variable heat demand.
- To enable plant to run at optimal loading to maximise efficiency.

In designing a store the following parameters need to be considered:

- Capacity—how much energy can be stored; this may vary over the year if the flow temperature from the production plant varies.
- Heat loss rate which will vary with the amount of heat in the store at any time.
- Maximum rate of charging.
- Maximum rate of discharging.

The size of the store needs to be optimised based on an hour-by-hour operating model and expected electricity and gas price variations.

In the future it is likely that price variations will occur not just on a daily basis but in relation to wind power outputs which may vary over several days or weeks. As a result, thermal storage is likely to become larger in relation to the plant capacity. This issue is discussed further in section 10.2.

It is important that the design of the store minimises mixing of flow and return water so that the stratification between hot layer and cool layer is maintained. To achieve this the flow rate of water into the store should be minimised by using two ports so that water only enters and leaves the store if it is being stored or discharged. In addition, the entering velocity needs to be reduced as far as possible using large diameter entry ports and a diffuser plate.

Where multiple tanks are required these are best connected in series rather than in parallel to minimise the amount of water in the mixing zones.

For further information on the design of thermal stores see [18].

6.13 District cooling production

As for a DH system the production plant for a DC system needs to have benefits over an individual building cooling plant to justify the cost of the DC network. These benefits typically include:

- Lower capital cost for chillers due to economies of scale.
- Higher efficiency of large-scale chillers.
- Use of low or zero GWP refrigerant leading to lower greenhouse gas emissions.
- Where low carbon low cost heat is available—the use of heat driven chillers.
- Free cooling from a source such as a river supplied with meltwater.
- Cooling obtained from a heat pump system that is also producing heat (see section 10.7).
- Use of evaporative cooling towers—not normally used in smaller buildings but results in higher efficiency.
- Chilled water or ice storage or other phase change material to optimise purchase of electricity to reduce costs.
- Less space required and less use of premium space such as roof top plant in prestige office blocks.

For more information on the design of DC systems see ASHRAE Guide to DC [19] and the Sustainable District Cooling Guidelines published by the IEA [20].

6.14 Thermal storage for DC

DC systems often use storage, especially to exploit the difference in electricity price between night and day periods. In countries where air-conditioning is essential there can be high peak electricity demands during the day which leads to high electricity prices in these peak demand periods. DC systems have used both ice storage and chilled water storage.

With ice storage the latent heat involved in the change of state from water to ice is exploited. As a result, the energy stored per unit volume is about four times that of chilled water. However, this benefit in terms of space saving needs to be compared with the lower efficiency of the chillers as they will need to produce cooling at a lower temperature and the cost of the chiller sized for ice making is also higher in £ kW^{-1} terms.

Most large DC stores use chilled water—the same medium as for the DC network using large cylindrical tanks. The volume is high compared to heating as the temperature difference between flow and return is around 10 °C at most. As the temperatures are low, concrete can be used rather than steel.

In the energy centre for Terminal 5 Heathrow there is a 3400 m^3 chilled water store constructed in concrete. This was designed to reduce the chiller capacity by meeting short-term demand peaks, to take advantage of differential pricing of electricity and to maximise the use of free cooling by operating the system at night

[21]. Much larger stores are found in the Middle East and Far East, for example the chilled water store at Kuala Lumpur airport is about 40 000 m^3 [22].

6.15 Energy centre building design

Once a DH or DC network has been installed in the ground there is no obvious visible indication that it exists. The energy centre will however be a new element of civic infrastructure and may have to be located near to built-up areas and so its aesthetic design is important. There are many examples where an energy centre has been designed to be visually attractive and enhance the local area in some way by providing community facilities. Often the energy centre is a focal point for visitors to the project and can provide an educational opportunity. Where there is a stack for flue gases or thermal stores, these have been opportunities for creative designs.

Some examples of interesting architectural approaches are:

- Kings Yard Energy Centre, Queen Elizabeth Olympic Park, London—winner of Stirling Prize for industrial architecture (https://www.archdaily.com/391885/olympic-energy-centres-john-mcaslan-partners)
- Islington, London, Bunhill Phase 1—thermal store cladding in timber and perforated screen of Phase 2 (https://www.islington.gov.uk/~/media/sharepoint-lists/public-records/energyservices/information/leaflets/20212022/20210923bunhill-heatandpowernetworkbrochure.pdf#:~:text=The%20original%20network%2C%20Bunhill%20phase%201%2C%20is%20po,electricity%20to%20heat%20buildings%20and%20provide%20hot%20water).
- Greenwich Peninsula, London—architect designed flue stack (https://www.steelconstruction.info/Greenwich_Peninsula_Low_Carbon_Energy_Centre,_London)
- Queens Quay Glasgow heat pump, Energy Centre glazed facade and flue/ventilation stack with lighting (https://www.cibsejournal.com/case-studies/a-new-era-for-heat-queens-quay-heat-pump/)
- Spittelau Energy from Waste CHP plant, Vienna—artist led design of façade (https://www.visitingvienna.com/footsteps/spittelau/)
- Viborg CHP station, Denmark—striking architectural design (https://en.wikipedia.org/wiki/Viborg_Power_Station)
- Amager Bakke Energy from Waste CHP plant, Denmark—artificial ski slope (https://worldarchitecture.org/architecture-news/emcpe/big-designed-copenhill-amager-bakke-wins-world-building-of-the-year-award-for-2021.html)

Although the aesthetic design will be important the main function of the energy centre is to provide space for the plant to be installed, operated, maintained and replaced in an efficient and safe manner. The Construction (Design and Management) Regulations, 2015 [23] require designers to carry out safety risk assessments and design to mitigate these risks. Some important considerations for the design team to work on together include:

- Access to install plant and to remove it at the end of its life without disrupting the heating/cooling service.
- Separation of plant areas to enable maintenance to take place whilst other plant remain in operation.
- Safe methods of access for maintenance work including replacement of smaller items of plant that may still be too large to manhandle easily; where possible plant requiring maintenance should be sited at low level not at height.
- Ventilation to provide combustion air, cooling air, emergency control of refrigerant leakage.
- Control of noise to both surrounding areas and internal areas.
- Storage for spare parts and workshop area.
- Office and control room together with welfare facilities (changing, washing, toilets, kitchen).
- If required for educational purposes, a meeting room with display boards or other community facility.
- The need for mechanical cooling in summer to maintain suitable working conditions and to provide conditions for electronic control systems.
- Embodied energy in the materials used for the energy centre construction.
- Designs to enable the energy cntre building to be maintained and kept clean including its immediate surroundings.
- Car parking for operatives and visitors with electric vehicle charging points.

The ideal aspiration is for the energy centre to fulfil all of its functional requirements well, be attractive visually and contribute in some way to the needs of the local community.

Bibliography

[1] Rogers and Mayhew 1992 *Engineering Thermodynamics Work and Heat Transfer* 4th edn (Longman Group) pp 64–5

[2] Chartered Institution of Building Services Engineers (CIBSE) 2013 *Combined Heat and Power for Buildings, Applications Manual AM12* (CIBSE)

[3] Michael M and Shapiro H 2006 *Fundamentals of Engineering Thermodynamics* 5th edn SI Units (Wiley) p 193

[4] Chartered Institution of Building Services Engineers (CIBSE) 2021 *Technical Memorandum TM65, Embodied Carbon in Building Services: A Calculation Methodology* (CIBSE)

[5] ISO 817: 2014/Amd 1: 2017 2017 *Refrigerants—Designation and Safety Classification* (ISO)

[6] Chartered Institution of Building Services Engineers (CIBSE) 2016 *Code of Practice for Surface Water Source Heat Pumps, CP2* (CIBSE)

[7] Chartered Institution of Building Services Engineers (CIBSE) 2019 *Code of Practice for Open-Loop Groundwater Source Heat Pumps, CP3* (CIBSE)

[8] Chartered Institution of Building Services Engineers (CIBSE) 2021 *Heat Pumps for Multi-Unit Residential Buildings, Applications Manual AM16* (CIBSE)

[9] Chartered Institution of Building Services Engineers (CIBSE) 2022 *Heat Pumps for Large Non-Domestic Buildings, Applications Manual AM17* (CIBSE)

[10] Energy Savings Trust, Gastec, AECOM, EA Technology 2009 *Insitu Monitoring of Efficiencies of Condensing Boilers and Use of Secondary Heating* (DECC)
[11] BEIS 2022 *Digest of UK Energy Statistics* (https://gov.uk/government/statistics/digest-of-uk-energy-statistics-dukes-2022)
[12] BRE 2023 *The Standard Assessment Procedure (SAP10.2)* (https://bregroup.com/sap/sap10/)
[13] BRE 2020 *Technical Note No. 3—Modelling Energy from Waste Facilities* (https://bregroup.com/sap/bre-technical-notes/)
[14] DECC 2014 *Timber Standard for Heat & Electricity: Woodfuel Used Under the Renewable Heat Incentive and Renewables Obligation* (DECC) (https://gov.uk/government/publications/timber-standard-for-heat-electricity)
[15] Chartered Institution of Building Services Engineers (CIBSE) 2014 *Biomass Heating, Applications Manual AM15* (CIBSE)
[16] Woods P *et al* 2005 A comparison of distributed CHP/DH with large-scale CHP/DH, Report 8DHC-05.01 *IEA District Heating and Cooling Project Annex VII* (https://www.iea-dhc.org/the-research/annexes/2002-2005-annex-vii/annex-vii-project-01)
[17] Chartered Institution of Building Services Engineers (CIBSE) 2016 *Heating, Guide B1* (CIBSE)
[18] Cabeza L (ed) 2015 *Advances in Thermal Energy Storage Systems* (Woodhead Publishing)
[19] ASHRAE 2019 *2019 ASHRAE District Cooling Guide (2nd edn) and Owner's Guide for Buildings Served by District Cooling* (ASHRAE)
[20] IEA 2020 *Sustainable District Cooling Guidelines* (https://www.iea-dhc.org/the-research/annexes/annex-xii/annex-xii-project-05)
[21] Overbye P and Woods P 2003 Chilled water store in heathrow airport *News from Danish Board of District Heating 4/2003* (https://studyres.com/doc/12926040/chilled-water-store-in-heathrow-airport)
[22] District Cooling in Malaysia 2018 *Malaysia Energy Information Hub* (https://egeda.ewg.apec.org/egeda/meeting/16WSpresentations/1–4.%20Malaysia.pdf)
[23] Construction (Design and Management) Regulations 2015 *Statutory Instrument 2015* No 51 (https://legislation.gov.uk/uksi/2015/51/introduction)

Paul Woods, MA MSc CEng FIMechE FEI FCIBSE

Chapter 7

Operating temperatures for DH

7.1 Introduction

The selection of operating temperatures for the DH network is a fundamental strategic design decision. The flow and return temperatures under peak design conditions will be used to determine the pipe sizes in the network and will influence the selection of pipe material. They will also impact on the efficiency, and in some cases the maximum output, of the central production plant and may influence the type of building interface used and the size of heat emitters and for existing buildings determine whether these need replacing.

In addition to the flow and return temperatures at peak, decisions need to be made about whether the flow temperature will vary through the year and in what relationship to demand or be kept constant.

There has been considerable work carried out under the IEA DH-C Programme, to evaluate ways in which operating temperatures can be reduced with low temperature schemes (<70 °C) being described as fourth generation [1]. Often these studies assume new build projects or buildings where the heat emitters can be replaced.

The selection of operating temperatures and especially the return temperature to be assumed will be governed by the buildings to be connected (whether existing or new) and the return temperatures cannot be defined solely by the DH designer without consideration of the operational requirements of the existing building or by dialogue with the designers of the new building.

The selection of operating temperatures needs to take account of a number of factors:

- The requirements of the **customer**—minimising disruption to the customer may be an important selling point for connecting existing buildings so the limitations of the existing space and water heating systems need to be considered. How the customers' heating systems are controlled and whether variable temperatures are acceptable needs to be investigated.

doi:10.1088/978-0-7503-5286-4ch7

- The requirements of the **heat network**—to reduce capital costs and pumping energy implies a large temperature difference between flow and return but to reduce heat losses a lower *mean* temperature is required.
- The requirements of the **heat production** plant—to increase output, efficiency and reduce costs lower flow and/or return temperatures may be needed.
- The requirements of **thermal storage**—storage volumes and hence costs will be lower with a large temperature difference.

It can be seen that there is a need for a balance as for a given return temperature higher flow temperatures reduce the capital cost and pumping energy of the heat network but may increase heat losses and the cost of heat production. Because there are many different opportunities for heat production it is inappropriate to set a standard temperature regime for all DH schemes as there will always be exceptions. However, as it is expected that heat pumps will dominate the heat supply in the future a practical operating temperature design strategy that is suitable for both heat pumps and existing buildings is set out in section 7.6.

7.2 Customer's systems

Traditionally in the UK space heating systems were designed for 82 °C flow 71 °C return which was set when coal-fired back boilers were in common use so as to avoid corrosion issues (the temperatures were expressed in °F, i.e.,180 °F flow, 160 °F return). Since about 2000, radiator selection tables have assumed 80 °C flow 60 °C return. Section 3.3 shows how the output of radiators reduces as the operating temperatures are lowered. It is usually found that there is a significant amount of oversizing of radiators due to innate conservative design approaches and in addition buildings may have been thermally upgraded since the radiators were installed. Experience of the building owners may also have established that operating with lower mean temperatures may still deliver sufficient comfort even under design conditions. When connecting an existing building designers will need to consider the potential for resetting design operating temperatures for space heating circuits which may mean also rebalancing of flow rates to achieve lower return temperatures. As the 82 °C flow temperature has been in place for a long time significantly higher flow temperatures would be seen as increasing safety risks.

A typical solution for most buildings would be to set 80 °C as the peak flow temperature but reduce the return temperature to 50 °C. This results in a reduction in heat output to 87% of the 80 °C/60 °C output which would be acceptable in most cases. To achieve this change will require resetting flow rates to 67% of the original design value which may mean introducing new balancing valves, new pressure independent thermostatic radiator valves, changing pumps and/or adding variable speed drives.

The requirement to operate at these temperatures is only present under design conditions and at times of part load the flow temperature can be reduced. This will

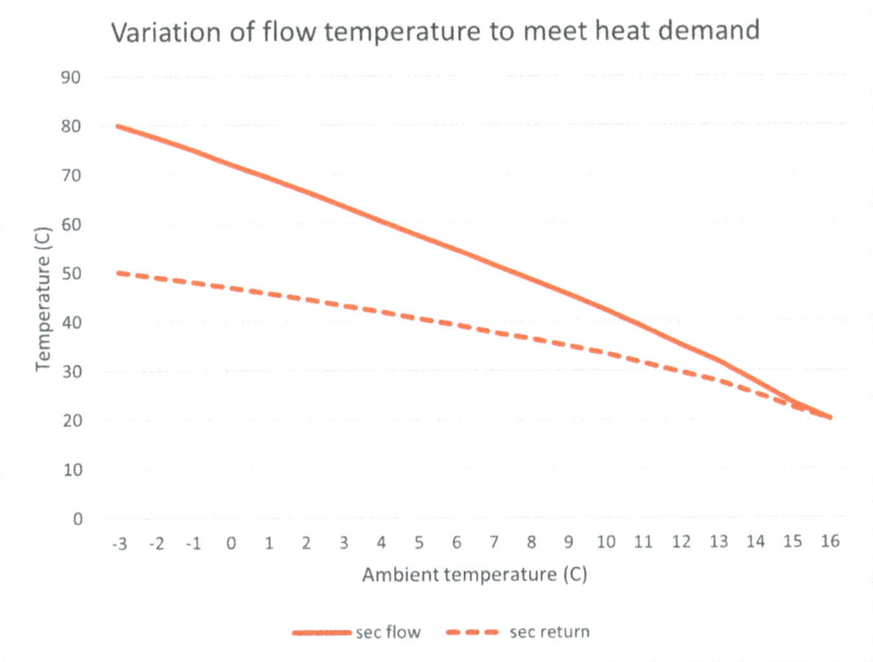

Figure 7.1. Minimum flow temperature to meet space heating demands.

be important where the heat production plant efficiency depends on the flow temperature, e.g., with a heat pump. To establish how the flow temperature can vary, the space heating demand has been estimated assuming that the peak demand occurs at −3 °C and no net heating demand is required above 16 °C (internal gains are usually sufficient to achieve comfort levels above this mean air temperature). The flow and return temperatures that provide a radiator output sufficient to meet this demand have been calculated and are shown on figure 7.1, plotted against mean daily outside air temperature.

For domestic hot water heating the required temperatures depend on whether storage is part of the system. Where storage is present it will be necessary to maintain the stored water temperature at 60 °C to prevent legionella bacterial growth. The supply temperature to the calorifiers will, therefore, need to be around 65 °C as a minimum. This will determine both the minimum network temperature that can be used and also the minimum output of the existing DHW system in summer. If this output is insufficient for the customer, then there will need to be modifications to the hot water system to provide sufficient capacity in the summer. This could involve exchanging calorifiers to instantaneous heat exchangers or a system where external heat exchangers are used in conjunction with a storage vessel (see figure 3.12) and in both cases the heat exchangers would be sized for the 65 °C minimum DH flow temperature.

7.3 Heat network

The cost of the heat network comprises: capital costs, pumping energy and heat losses.

Reducing capital costs
DH networks transfer energy by pumping water from the production plant to the customers. It follows that minimising the volume of water to be transported should lead to a lower cost system.

To minimise the volume of water (V) needed to transfer a given amount of energy Q the temperature difference between flow and return (ΔT) needs to be maximised. V is inversely proportional to the temperature difference:

$$\Delta T = T_f - T_r$$
$$Q = c\,\rho\,V\,\Delta T \text{ or}$$
$$V = Q/(c\,\rho\,\Delta T)$$

where

c = specific heat capacity (kJ kgK^{-1})
ρ = density (kg m^{-3})

This principle applies to the peak demand condition as it is the peak flow rate that governs the size and hence the cost of the pipes -a key component in the economics of the scheme.

However, the impact of the ΔT and hence volume flow on cost is less than might be expected. There are two reasons for this:

- The pipe diameter is related to the square root of the volume flow for the same velocity (and smaller pipes will typically be sized optimally at lower velocities—see section 5.5.6).
- The cost of the network is not related linearly to the diameter due to a high fixed cost element.

An approximate relationship between capital cost and diameter was given in section 5.5 and is shown in figure 7.2.

For example:

Consider a pipeline carrying 2000 kW with a ΔT of 30 °C (80 °C/50 °C) An optimal pipe size would be 100 mm NB with a velocity of 1.8 m s^{-1}.

If the ΔT is doubled to 60 °C (95 °C/35 °C) the flow rate would halve and the optimal size is then 65 mm with a velocity of 2.1 m s^{-1}.

From figure 7.2 the capital cost reduction is from £828 m^{-1} to £704 m^{-1} a reduction of 15% and the total annualised cost (calculated as in section 5.5.5) reduces from £69 m^{-1} p.a. to £60 m^{-1} p.a. a reduction of 13%. It can be seen that the cost of distribution is relatively insensitive to changes in ΔT as even a doubling of ΔT and halving of flow rate reduces total network costs by only 13%.

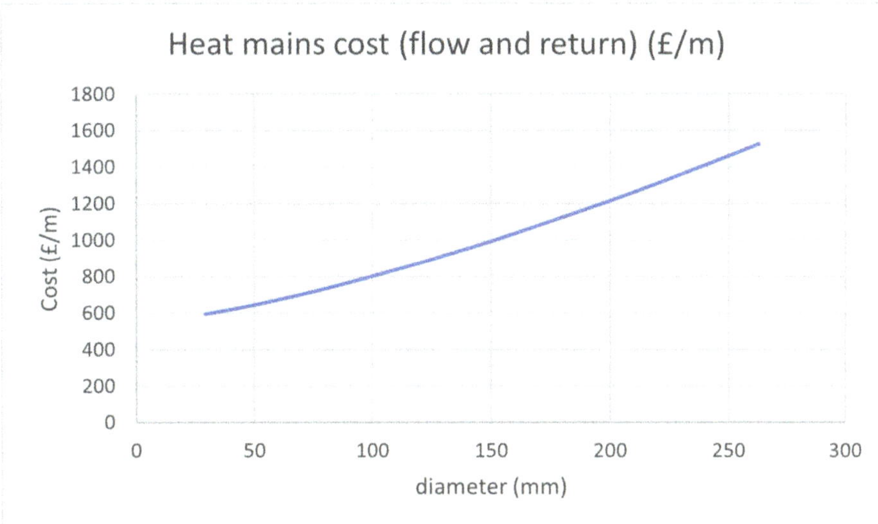

Figure 7.2. Capital cost of flow and return pipes against internal diameter.

Pumping energy

Pumping energy is related to the volume of water and the required pressure difference (see section 5.3). Pumping energy will be reduced by using a larger ΔT and hence lower volumes.

Heat losses

One of the objectives in using lower operating temperatures in so called 4G systems is to reduce heat losses.

Heat losses are related to the difference between the mean temperature of the network water, i.e., $(T_f + T_r)/2$ and the ambient temperature. So reducing the mean temperature for the same ΔT will lead to lower heat losses. However, if the ΔT is smaller the flow rates are greater and larger pipes will be needed which will have a higher heat loss due to their greater surface area. Pumping energy also changes as a result of the changes to flow rates and diameters

To compare the heat losses for different temperature assumptions it is necessary to firstly calculate an optimal pipe size for each flow and return temperature and then assess the heat losses and pumping energy.

The calculation will vary depending on the network and its total length in relation to the heat demand as total pressure may act as a constraint on pipe sizing. However, as an illustration table 7.1 has been calculated on a per m basis, using 3000 kW heat rate and economic optimal pipe sizing as described in section 5.5.

Taking Case A (85 °C/55 °C) as the base case the table shows:

Case B (95 °C/55 °C) increases the flow temperature which increases the ΔT, leading to a smaller pipe and lower capital costs. The heat losses reduce as even though the flow temperature is higher the smaller pipe has a lower heat loss

Table 7.1. Cost comparison for a range of operating temperatures for a 3000 kW heat rate.

Case	Flow temp	Return temp	ΔT	Pipe size NB	Velocity	Heat loss (F and R)	Pumping energy kWh m^{-1}	Cost of heat loss	Cost of pumping energy	Total energy cost	Annualised capital cost	Total annualised cost	Percentage from base case
	°C	°C		mm	m s^{-1}	kWh m^{-1} p.a.	p.a.	£ m^{-1} p.a.	£ m^{-1} p.a.	£ m^{-1} p.a.	£ m^{-1} p.a.	£ m^{-1} p.a.	%
A	85	55	30	125	1.78	251.1	27.1	12.56	4.07	16.63	57.77	74.40	—
B	95	55	40	100	2.04	229.7	34.7	11.49	5.21	16.7	51.74	68.44	−8%
C	75	55	20	150	1.82	256.6	33.7	12.83	5.06	17.89	64.91	82.80	11%
D	75	45	30	125	1.78	206.1	27.1	10.30	4.07	14.37	57.77	72.14	−3%
E	65	45	20	150	1.82	205.6	33.7	10.28	5.06	15.34	64.91	80.25	8%
F	85	45	40	100	2.04	190.6	34.7	9.53	5.21	14.74	51.74	66.48	−11%

Energy costs are based on 5 p kWh^{-1} for heat and 15 p kWh^{-1} for electricity for pumping, i.e., this assumes a heat pump with a CoP of 3 for calculating the cost of heat, a load factor of 20% has been assumed when calculating annual pumping energy from the peak pressure drop.

coefficient. The pumping energy is higher in this case as although the volume flow is lower the velocity is higher and the diameter smaller resulting in a higher pressure drop which more than offsets the benefit of lower volume. The total annualised cost is 8% lower so raising the flow temperature to improve the ΔT lowers the total cost of the network.

Case C (75 °C/55 °C) has a lower flow temperature which reduces the heat loss from the flow pipe but the smaller ΔT means a larger pipe which offsets this so that heat losses are slightly higher. Pumping energy is higher as the volume is 50% more than the base case, even though the velocity is slightly lower. So the net effect is an increase in operational energy costs. The larger pipe size means the capital cost for this case is higher and so the total cost is 11% higher than the base case.

Case D (75 °C/45 °C) uses the same temperature difference as the base case so the pipe size and pumping energy are the same as the base case. Heat losses are lower by about 18% as both flow and return temperatures are lower. Total cost reduces by 3%.

Case E (65 °C/45 °C) uses a smaller ΔT (as Case C) but at a lower temperature level reducing heat losses by 18%. The larger pipe size means capital cost is higher and the total cost is 8% higher than the base case.

Case F (85 °C/45 °C) reduces the return temperature to improve the ΔT and results in a smaller pipe size as for Case B. Heat losses reduce due to the smaller pipe size and lower return temperature. Pumping energy is slightly higher as for Case B. The smaller pipe size leads to a lower capital cost and overall the cost reduction is 11%. Reducing the return temperature is beneficial for the network and has a bigger benefit than increasing the flow temperature as in case B.

The conclusion from the above is that (assuming constant temperatures through the year) lowering flow and return temperatures with no change in ΔT results in a small reduction in total cost but increasing the ΔT is the most beneficial approach as this will lead to a reduction in total cost with a 10 °C reduction in return temperature having more impact than a 10 °C increase in flow temperature (compare Cases B and F with Case A). If the flow temperature is reduced with the same return temperature a smaller ΔT results which leads to larger pipe size, no benefit in lower heat losses and higher overall costs (compare Case C with Case A).

Comparing Case F 85 °C/45 °C and E 65 °C/45 °C is also instructive. The lower flow temperature of Case E might be expected to reduce heat losses and result in lower overall costs. In fact, the reverse is found, heat losses are higher due to the larger pipe size and Case E is 21% more costly than Case F.

The cost variation from the base case for these examples is less than ±11% which, given that the network cost is only one component of the total cost of the project, indicates that the operating temperatures selected are not too critical. On Page 45 of the IEA report [1] it is concluded that the use of low temperatures in reducing heat losses is less significant than the benefit of lower heat production cost, which is discussed in section 7.4 below.

The above analysis assumes that there is no difference in cost for a network designed for 65 °C or 95 °C in terms of capital cost per m installed. In practice there may be some cost benefit in using lower temperatures as there will be a reduced need for provision for expansion (see section 5.7) and where plastic pipes are used the lifetime of the system is temperature dependent (see section 5.6.2). Use of temperatures over 100 °C is usually avoided with new networks unless there are special reasons as there will be additional costs for expansion provision and the use of double isolation valves at building connections.

7.4 Heat production plant and selection of temperatures

7.4.1 Gas-engine CHP

Heat is produced from exhaust gases, oil cooler and jacket cooling and typically temperatures up to 90 °C can be achieved with no loss of performance. Recovery of some or all intercooler heat is possible with low return temperatures of 45 °C or lower. So for this heat source a heat network operating at 90 °C flow at peak will be optimal given the conclusions of section 7.3.

7.4.2 Heat pumps

For heat pumps, the coefficient of performance (CoP) (useful heat per unit of electricity) is very dependent on the flow temperature (supply temperature) as shown in figure 7.3 and this can have a major influence on selection of operating

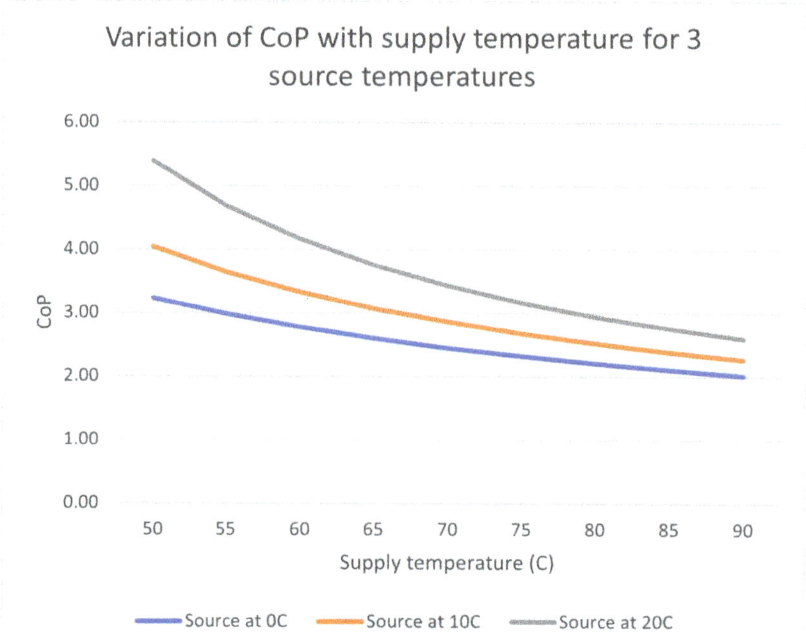

Figure 7.3. Variation of CoP for a heat pump assuming Carnot efficiency of 50%.

temperatures. In addition, the maximum flow temperature that can be achieved varies with the refrigerant used. There is also some impact on CoP with varying return temperatures, however, there is also a maximum temperature difference across the heat pump so there may be no benefit in reducing return temperatures to improve heat pump performance below a certain limit. Where CO_2 is used as the refrigerant the return temperature needs to be lower than 40 °C.

7.4.3 Steam turbine

The temperature at which steam is extracted from the turbine directly influences the z-factor. So typically a z-factor of 7 might be achieved at 110 °C, and a z-factor of 10 at 90 °C. For large-scale power plants, more than one extraction point may be used, e.g., taking the DH water from 55 °C to 95 °C in two stages of heating.

7.4.4 Boilers

The efficiency of boilers depends mainly on the return temperatures as low return temperatures will enable the flue gases to be cooled to a lower temperature. If sufficiently low the water vapour in the flue gas condenses releasing more heat. The impact is most significant with biomass or waste streams where there is a higher proportion of water vapour in the flue gas. Condensing gas boilers are normally specified as higher efficiencies can be obtained with return temperatures below 50 °C.

7.4.5 Heat recovery from industry

Where there is heat recovery from a waste heat source it will be the return temperature that will govern the amount of heat that can be recovered and so the return temperature has a very direct impact on the viability of the heat source. The flow temperature will be determined by the temperature of the heat source. If the network temperature is higher then heat pumps would be needed which would add to the cost so using a lower flow temperature in this situation would be important to avoid the cost of a heat pump.

7.4.6 Deep geothermal

If the return temperature is lower more heat can be extracted from the borehole and as the main cost is the cost of the borehole itself the return temperature has an impact on the size of the resource and the economics of the scheme. But if the flow temperature selected is too high and heat pumps have to be used then this will add to the capex and opex of the project.

7.5 Variable flow temperatures

7.5.1 Operating strategies

The network will be operating only occasionally at peak conditions and so efficient operation at part load is key.

The network operation is largely governed by the decisions made on the customer's heating systems and the substations. The substations will typically use two-port control valves thereby creating a variable volume system.

Where agreed with the customer a variable temperature secondary system may be used and if this is the case for all customers then the DH network can also operate as a variable flow temperature system.

Depending on the type of DHW heating installed at the customers the minimum DH flow temperature at the customers will be limited to ca. 65 °C where storage is used and 55 °C where instantaneous hot water is provided (see section 3.3.3).

In section 7.3 it was shown that space heating demands can be met by reducing secondary flow temperatures from 80 °C at peak at −3 °C outside air temperature to around 60 °C for a +5 °C outside air temperature.

Within these constraints there are therefore two options:

(A) Operate as variable volume flow and constant flow temperature.

(B) Operate as variable volume flow but with variable flow temperature down to a minimum which allows for DHW production.

Option A:

Lower volume flow rates at part load results in two benefits:

- Reduced pumping energy needed as both the volume and the pressure drop falls:

  pump energy = volume flow x head required/pump efficiency
- Lower heat losses from the return:

  as the volume reduces the return temperature falls assuming a two-port control valve is used within the building (see figure 3.7).

Option B:

The second option is to lower the flow temperature at part load. This will minimise heat losses but pumping energy will be higher. Also, the flow temperature can normally be reduced only to about 65 °C because some customers may require this for DHW production. With a lower flow temperature, return temperatures will be slightly higher than for the constant temperature case (see section 3.2.4).

For Option B there is a trade-off between lowering heat losses and increasing pump energy. Therefore, Option B is likely to be of most interest where the cost of heat production is high and the required pressure difference at the pumps is relatively low (e.g., a small scheme using a heat pump). Conversely, where the cost of heat production is low and pressure difference is high (e.g., an energy from waste plant with a transmission main and an extensive network) it may be better to maintain a constant flow temperature.

A network simulation that calculates the pumping energy and heat losses on an hour-by-hour basis for different operating temperature assumptions is required to establish the best strategy.

Reducing the temperature at part load will also extend the life of plastic pipes if used.

7.5.2 Heat production plant

The above discussion assumed that there is no impact on the cost of heat production by varying temperatures. Whilst this is the case for say gas-engine CHP it is not the case for heat pumps which are likely to become dominant in the future. A lower flow temperature will improve the CoP of the heat pump. This means that Option B will be more effective, i.e., reduce flow temperatures in relation to load and use variable volume control to achieve lower return temperatures.

However, the peak load would typically be met with supplementary gas boilers and if these are in series with the heat pump the impact on the CoP of higher peak temperatures is reduced or eliminated. The heat pump can continue to operate at ca. 70 °C with the rise in temperature required in cold periods achieved by operating the peak boiler plant. For a typical load duration curve (see section 2.1.6), the primary source can be sized for 50% of the peak and the peaking plant will then deliver at most 15% of the annual demand.

Figure 7.4 shows how a typical heat demand varies with external air temperature. A primary heat source sized at 50% of peak would meet all of the demand above a mean temperature of 8 °C.

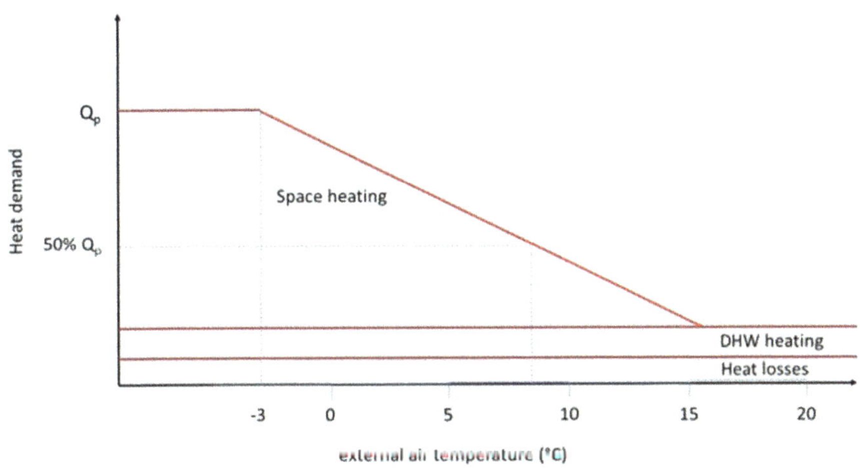

Figure 7.4. Typical load variation with external air temperature.

Figure 7.5 shows how peak boilers in series with the primary heat source could be used to raise the temperature from 70 °C at 8 °C ambient temperature to a maximum of 85 °C at −3 °C. Also plotted on figure 7.5 is the minimum secondary space heating temperatures required to meet the space heating demand, taken from figure 7.1. It can be seen that there is considerable margin between the DH flow temperature and the secondary flow temperature. This means that a primary heat source could still operate at 70 °C even if it was sized at more than 50% of peak load.

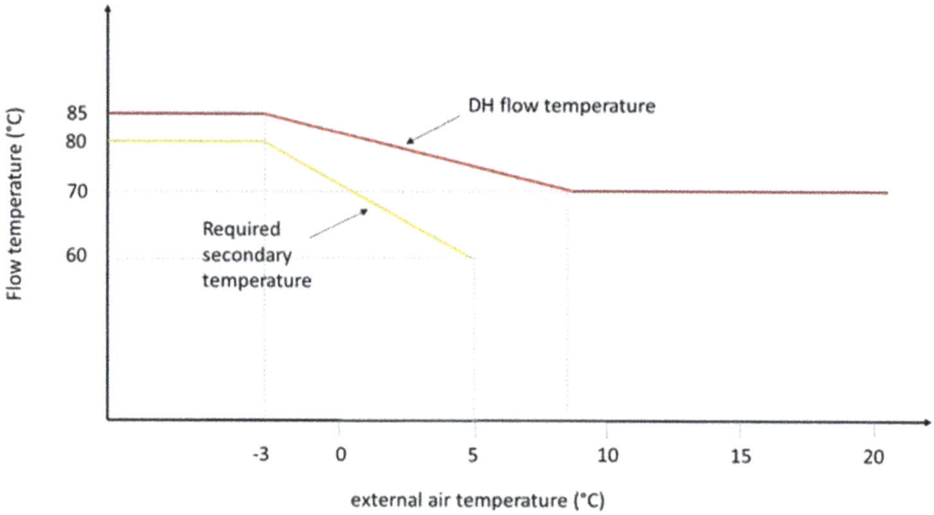

Figure 7.5. Minimum operating flow temperature in relation to external air temperature.

7.5.3 Thermal storage

The size of the thermal store is determined by the temperature difference between flow and return. A large temperature difference would, therefore, be preferred. However, if the flow temperature is reduced at part load the storage capacity will also be reduced. If return temperatures under part load and summer conditions can be lowered then this will enable the stored energy to be similar to peak conditions throughout the year.

7.6 Summary

From the above discussion the following key principles can be stated:

1. Reducing the return temperatures from the customers as far as practical will normally be worthwhile as this will reduce network capital costs, heat losses and pumping energy. Also, low return temperatures will maximise the potential heat available from waste heat recovery and deep geothermal sources.
2. For existing buildings, to avoid replacing the heating emitters circuits should be rebalanced to achieve lower return temperatures of about 50 °C or lower. For new buildings, selecting low return temperatures in the range 30 °C to 40 °C is likely to involve only a small additional cost in heat emitters. Return temperatures from DHW heating can be 20 °C or lower as typically the cold feed temperature will be 10 °C or less under winter conditions.
3. For a given return temperature the total heat network cost will be lower with a higher flow temperature as the smaller pipe reduces capital cost and can lead to lower pumping energy and lower heat losses.

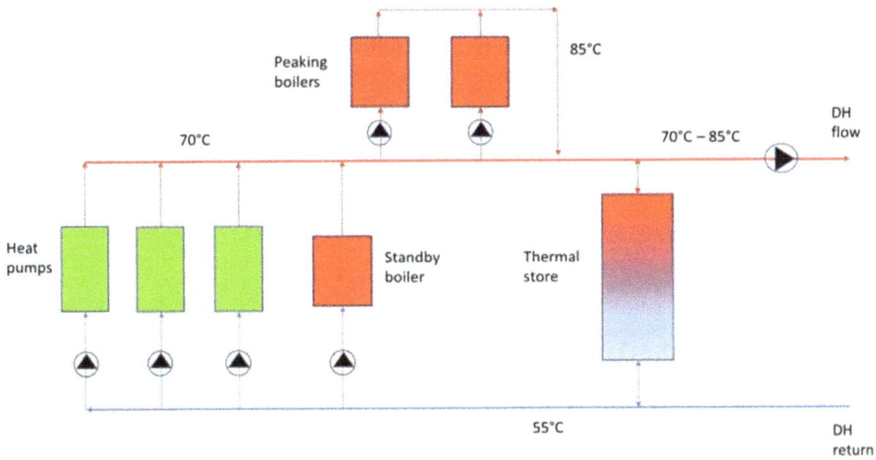

Figure 7.6. Typical simplified schematic with peak boilers in series with heat pumps.

4. However, where heat pumps are used the flow temperature has a strong influence on efficiency such that a lower flow temperature would be preferred to reduce operational costs and emissions.
5. To maintain domestic hot water storage above 60 °C a minimum flow temperature of 65 °C is required at the customer so 70 °C is assumed to be needed at the heat production plant to allow for network losses.
6. An optimal strategy for a heat pump based DH system would be to operate the network at 70 °C for most of the year with the flow temperature being raised for ambient temperatures below about 8 °C, rising to 85 °C at peak demand (see figure 7.5). This will enable the minimum secondary space heating temperature as in figure 7.1 to be met.
7. For most projects using peak boilers this temperature rise can be achieved through operating boilers in series with the heat pump, so the efficiency of the heat pump is unaffected. The disadvantage of this arrangement is that condensing conditions would not be achieved. A temperature rise from 55 °C to 85 °C would match a heat pump operating at 70 °C and sized at 50% of the peak demand (see figure 7.6). Typically, with this approach, about 15% of the demand will be met by peak boilers.
8. A maximum temperature of 85 °C will enable direct connection where appropriate without a significant increase in the risk of scalding. It will also extend the life of plastic pipe systems (see 5.6.2).

Bibliography

[1] Averfalk H *et al* 2021 ed K Lygnerud and S Werner *Low Temperature District Heating Implementation Guidebook, Final Report of IEA/DHC Annex TS2* (IEA)

IOP Publishing

An Introduction to District Heating and Cooling
Low carbon energy for buildings
Paul Woods, MA MSc CEng FIMechE FEI FCIBSE

Chapter 8

Some operational aspects

8.1 Optimisation of operation of production plant

One of the advantages of DH/C is that the production plant can be more complex than the typical system for individual buildings. When using more than one type of heat production and thermal storage there is considerable flexibility as to how the plant can be operated.

For example, the low carbon, low running-cost plant, such as a gas-engine CHP or heat pump will usually operate at base load with gas boilers used at times of peak. The CHP plant would be used to fill the thermal store during the day so that it can be discharged in the evening and at night when the CHP is normally shut down due to low electricity prices. In contrast, the heat pump would operate mainly at night and fill the thermal store so that it can be discharged the following day and allow the heat pump to be shut down at times of peak electricity prices. A system with both a heat pump and CHP would operate the heat pump more at night and the CHP more in the day.

It is often the case that a balance needs to be struck between operating to minimise operating costs and operating to minimise CO_2 emissions. For example, it may be more economic to switch to using boilers at night instead of CHP, but lower CO_2 emissions would be obtained by operating the CHP through the night.

The above examples are a result of electricity prices varying between night and day. In some contracts there will be additional charges for drawing electricity at times of peak grid demand (Triad charges) or additional payments for increasing electricity demand when required by the National Grid. With more complex contracts it is likely that more variable prices will be seen, and in the future prices may be related to the real-time generation output from wind and solar. The ability to flex the production of electricity from CHP or the demand for electricity for heat pumps by using thermal storage will bring additional financial benefits to the operator (see section 10.2.4).

doi:10.1088/978-0-7503-5286-4ch8

In the future, the optimal operating strategy will need to be determined on a day ahead basis and not hardwired into a control logic. An automatic optimisation system will need to use predictive control where the heat demand is predicted based on weather forecasts and electricity prices and CO_2 emission factors are also predicted based on the expected levels of renewable energy with the optimal operating strategy calculated accordingly.

Optimising the operation over the year may involve varying the flow temperature to improve overall energy efficiency. Where heat pumps are used, operating with as low a flow temperature as possible at any time will be important (see section 7.5).

8.2 Water treatment

An important operating requirement is to maintain the water quality of the DH/C system so that internal corrosion does not reduce the planned life of the system and performance of heat substations and HIUs is maintained. This work starts in the design phase and a fully operational treatment system will need to be in place from the start including when commissioning the system.

Water treatment can be considered as two linked aspects: physical and chemical.

Physical treatment involves removing particulate matter in the water which is typically achieved using in-line and sidestream filters (often with magnetic filters to retain magnetite) and removing oxygen by vacuum degassing.

Conventional chemical treatment involves the addition of treatment chemicals which are designed to:

- Raise the pH to between 9.5 and 10 which slows corrosion of steel
- Remove oxygen that is dissolved in the water
- Stop particles from coalescing

More recently, an electrochemical approach has been adopted which relies on initially filling the system with demineralised water and maintaining this quality through treatment of the make-up water using an ion exchange resin filling device. The process removes the salts, minerals and carbon dioxide and provides a controlled fill with a pH of 8.2. The oxygen removal is carried out by the process of electrochemistry using magnesium anode technology. The dissolved oxygen is consumed by the magnesium anode, this releases magnesium hydroxide which controls the pH level, conductivity level and dissolved oxygen levels. Any future top up or replacement water is provided by the filling device using ion exchange resin. Other processes may be used provided there is compliance with the German standard VDI 2035 Parts 1 and 2.

For either option, regular monitoring of the water quality is important against a number of key indicators with corrective action taken when required.

The most important parameter to be monitored is the make-up water volume. Any system will lose small amounts of water at pumps, valves, vents and from maintenance work and this should be monitored weekly as any rise in this rate could indicate a leak in the system which will need to be located and repaired before control of water quality is lost.

Further details of water treatment methods are given in Objective 6.3 and Annex E of the CIBSE CP1.2 [1].

8.3 Metering and billing

8.3.1 Introduction

It is now a requirement for new buildings and dwellings to have heat meters installed under the Heat Network (Metering and Billing) Regulations 2014 [2]. Retrofitting of heat meters is also required where it can be shown to be cost-effective and technically feasible. Operators of heat networks need to provide notification of the network to the Office for Product Safety and Standards. Heat is sold to customers using a thermal energy meter which measures the volumetric flow rate and simultaneously the flow and return temperatures. An integrator (electronic calculator) takes these inputs and calculates and stores the thermal energy consumed in accordance with EN1434 [3].

Battery and mains powered meters are available; battery powered meters are preferred as they are generally cheaper to purchase and install but more importantly they are always on and are not affected by the state of the power supply to the property. Suitable maintenance arrangements need to be in place to replace the batteries, or exchange the meters, in due time, which may coincide with the need for recalibration.

Typically, meters will have a physical display showing date, time, kWh to date, calculated kW for the most recent period, volume flow rate for the most recent period and flow and return temperatures. The meter will also interface with an automatic meter reading (AMR) system.

8.3.2 Meter accuracy

Heat meters are manufactured and tested to comply with the Measurements Instrument Directive (MID). Unlike gas, electricity and other meters the MID was not transposed into UK law for heat meters. However, the current heat meter regulations require the heat meters to be accurate and the regulators have stated that they will take MID approval as meeting this requirement. It is likely that MID compliance will be a requirement going forward in revised legislation.

The most reliable volumetric measurement technique is to use ultrasonics to measure velocity through a tube. It is important that the manufacturer's recommendations are followed closely with regard to the installation for the required accuracy to be obtained and so that the MID compliance is not compromised.

8.3.3 Automatic meter reading systems

Modern heat network automatic meter reading (AMR) systems for dwelling meters typically use Mbus [4], for remote collection of meter data. This is the most commonly used protocol in the industry.

The Mbus meter is connected to an AMR system either by data cable bus or wirelessly. There is an additional standard for wireless Mbus, which incorporates a security layer, called OMS [5]. Generally wired Mbus is used in new build where installing new cables is practical and can be done at a reasonable cost. Wireless Mbus is usually reserved for retrofits.

The two most common AMR systems for new build or deep retrofit of a building are:

(A) To feed the readings back, wired or wirelessly, to a central building gateway with a connection to the internet usually via a SIM card or ethernet connection. The gateway requires a mains power supply.
(B) Connect the heat meter to a smart metering pay as you go (PAYG) unit using wired Mbus. The smart metering unit is usually equipped with a SIM card for onward data transmission over the mobile network.

Usually, the building owner will specify if they wish to use a smart metering system with PAYG functionality.

The advantages of smart metering systems are that they can be used to limit debt on the scheme by using the PAYG feature, and they help the residents manage their budgets.

Wired Mbus systems are also available with remote disconnection facilities.

The two options for disconnection are;

(A) A two port normally closed motorised valve (it is best to avoid valve actuators with long closing times).
(B) An electronic HIU with built in facility.

With the rising cost of energy, it is likely that more features will be added to smart metering systems that will allow the restriction of demand and or minimum temperature as a measure to protect residents with vulnerabilities.

It is not usually recommended that the meters are connected to the BEMS as this tends to be more expensive and complicated to install and the BEMS prioritise control over data collection, so reliability is not as high, often leading to more customer complaints.

It is common for client bodies to contract with a specialist metering company to provide the metering and billing service either separately from an operating and maintenance contract or as part of this contract. Ideally the metering and billing service provider should be appointed early in the project to advise and support during the design period and to allow time to prepare for first occupancy.

8.3.4 Use of meter data

Traditionally, heat meters are used to provide data for billing purposes using the cumulative energy register (kWh). Increasingly, the data from the heat meter is being used for monitoring of the system especially the flow and return temperatures at the point of connection to the customer.

A heat meter will be measuring volume flow as well as temperatures and so a key indicator of performance is the volume flow per unit of heat energy delivered. An abnormally high figure could indicate that the controls are not working as intended, for example with excessive bypass flows (if downstream of the meter), or a control valve stuck open or an unstable control valve (possibly with an incorrect set point) which is changing from fully open to fully closed regularly (hunting). The volume to energy ratio may also indicate poorly balanced radiators or under floor heating systems.

Modern heat meter systems can be set up to monitor data and provide automatic reports on an exception basis. The use of datalogging and automated analysis will enable faults to be picked up quickly that have an impact on the efficiency of the network (see figure 8.1).

Fla Ro	2.34	1.22	0.86	0.04	0.02	0.03	0.01
Fla	0.46	0.51	0.62	1.14	0.17	0.20	0.01
Fla Ro	0.44	0.45	0.78	0.67	0.16	0.09	0.02
Fla	0.34	0.55	0.63	0.49	0.15	0.10	0.07
Fla	0.31	0.39	0.47	0.33	0.16	0.17	0.06
Fla	0.16	1.00	0.04	0.22	0.14	0.16	0.05
Fla Ro	0.39	0.53	0.34	0.28	0.06	0.03	0.00
Fla Ro	0.10	0.09	0.13	0.34	0.08	0.33	0.00
Fla	0.29	0.22	0.49	?	0.02	0.01	0.00
Fla Ro	0.09	0.13	0.41	0.20	0.01	0.01	0.01
Fla	0.09	0.14	0.20	0.15	0.03	0.09	0.04
Fla	0.14	0.13	0.13	0.11	0.04	0.04	0.04
Fla	0.16	0.18	0.10	0.09	0.03	0.03	0.01
Fla	0.08	0.13	0.10	0.07	0.09	0.08	0.01

Figure 8.1. Weekly Apartment Volume to Energy Ratio Tracker using domestic heat meter data used in proactive maintenance (showing ratios for 14 flats over 7 weeks). Reproduced with permission from Switch2 Energy Ltd.

Meter reading frequencies

In the UK, half hourly meter reading has become the standard adopted by the electricity industry and is also available on most AMR systems on heat networks, this data can be presented to residents so they can monitor the usage of their heat and control settings. Operators also have more detailed data to work with and build daily half hourly demand profiles.

AMR systems now allow for more frequent read frequencies up to 2 min or less. However, there is a cost associated with collecting and storing data and many functions such as billing and monitoring return temperatures can be adequately accomplished using daily reads (a read collected usually at around midnight each day) where the AMR system does not support half hourly reading. It is also possible to read the data more frequently to help with identifying specific issues with HIU's operating modes. Increasing frequency of data collection has implications for the meter's battery life.

HIUs have evolved and can now be specified with systems that can be used to collect meter readings as well as providing two-way communication between a central control system and the substation or domestic HIU by which means some faults can be corrected remotely or set points altered. One of the benefits of this type of HIU communication system is speed and accuracy of installation, commissioning and verification.

Minimum requirements for metering are given in Objective 6.2 of CIBSE CP1.2 [1].

8.4 Customer service standards and regulations

The primary concern of the operator of a DH/C scheme will be to fulfil the contracts for supply of heating or cooling. Standard forms of heat supply contracts, part of the suite of contracts known as SOMS, have been produced by Lux Nova and published by Triple Point, the Heat Networks Investment Project partner organization [6].

The voluntary scheme The Heat Trust was set up to provide a greater level of customer protection and hence to assist the development of heat networks [7]. The contractual requirements in the Scheme Rules need to be incorporated into the heat supply contracts used in projects which have joined the scheme.

In 2018 the Competition and Markets Authority published a review of the heat network market [8] and made recommendations to introduce regulations to provide greater customer protection. These recommendations are now being taken forward in the Energy Security Bill, see the Policy Statement on the Heat Network Market Framework [9].

Bibliography

[1] CIBSE Heat Networks 2020 *Code of Practice for the UK* CP1.2 (CIBSE)
[2] HMG Heat Networks (Metering and Billing) Regulations 2014 as amended 2015 and 2020, HMG, 2020 (https://gov.uk/guidance/heat-networks)
[3] EN1434 2015 *Thermal Energy Meters* Parts 1–6 (European Committee for Standardization)
[4] Mbus *The Standard for Remote Reading of Smart Meters* (https://m-bus.com/)
[5] OMS *Open Metering System* (https://oms-group.org/en/)
[6] Triple Point 2023 *Template Contracts, Triple Point Heat Networks* (https://tp-heatnetworks.org/heat-contract-templates/)
[7] Heat Trust 2023 *Scheme Rules* (https://heattrust.org/the-scheme-rules)

[8] CMA 2018 *Heat Networks Market Study Final Report*, Competition and Markets Authority (https://gov.uk/cma-cases/heat-networks-market-study)
[9] HMG 2023 *Energy Bill Policy Statement, Heat Network Market Framework* (https://gov.uk/government/publications/energy-security-bill-policy-statements-and-draft-regulations)

Part III

Future opportunities and the case for DH/C

IOP Publishing

An Introduction to District Heating and Cooling
Low carbon energy for buildings
Paul Woods, MA MSc CEng FIMechE FEI FCIBSE

Chapter 9

Comparing DH/C with individual heating/cooling systems

9.1 Introduction

The concept of the 'energy trilemma' has been developed by the World Energy Council [1] which has published a trilemma index for over 100 countries annually since 2010. There are three related challenges illustrated in figure 9.1:

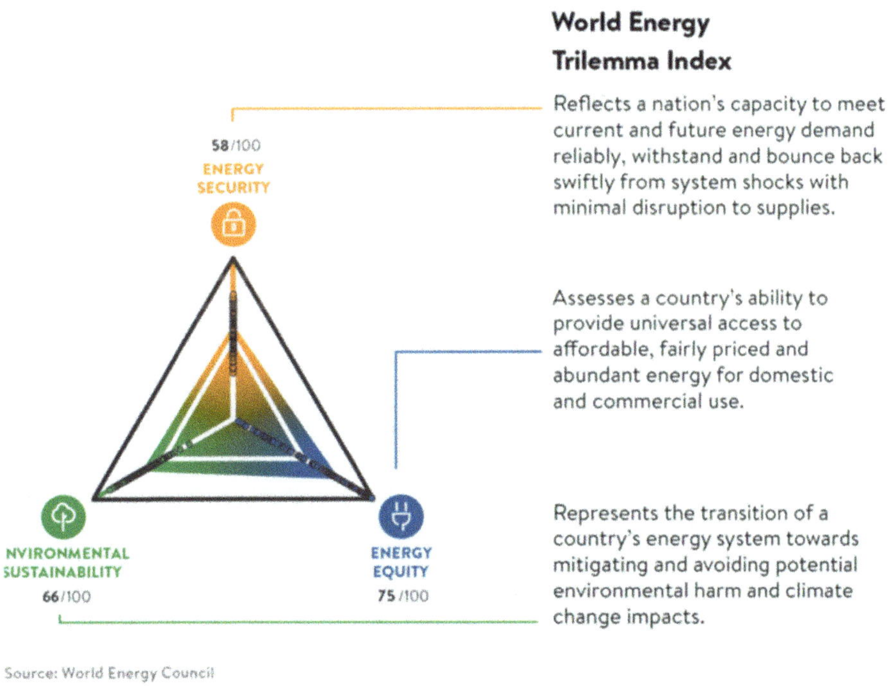

Figure 9.1. The energy trilemma (World Energy Council). Reproduced with permission from World Energy Council.

doi:10.1088/978-0-7503-5286-4ch9

- Environmental—mitigation of climate change impacts by minimising greenhouse gas emissions whilst protecting the environment and enhancing biodiversity
- Economic—providing energy at an affordable cost to all
- Security of supply—ensuring a country's economy is protected from disruption from external shocks to energy supply

These three challenges are discussed in this chapter in relation to DH/C.

In addition, there are a number of other aspects of DH/C which differ from say installing individual heat pumps within buildings:

- The nature of the market, how customers decide on energy options and where the technical risk lies.
- The customer experience, how difficult are energy options to access and what level of consumer protection is available.
- The financing of the energy options, how is finance provided and at what cost.

Most DH/C projects will have an element of public sector finance or ownership so when considering the case for DH/C an assessment using the 5-case model as recommended by HM Treasury [2] is often required. This comprises:

- The Strategic Case (what is the case for change?)
- The Economic Case (does it provide best value to society?)
- The Commercial Case (is it attractive to the market?)
- The Financial Case (is it affordable and how is it funded?)
- The Management Case (how it will be delivered and risks managed?)

The chapter concludes by discussing the main points that typically emerge when developing the case for DH/C projects.

9.2 Environmental

9.2.1 CO_2 emissions

A major driver for implementing DH/C projects will be to reduce the CO_2 emissions associated with the supply of heating and cooling.

For DH systems, the overall CO_2 content per kWh of heat delivered needs to be calculated.

This will be based on the operational energy model and will summate the CO_2 emissions associated with combustion of fuels at the energy centre for both CHP plant and boilers, the use of electricity for heat pumps, pumping and ancillary equipment and the displacement of CO_2 by the electricity generated by CHP. By dividing these total net emissions by the heat delivered to customers, the impact of heat losses from the network will be taken into account.

This figure expressed as kg CO_2 per kWh heat delivered p.a. can be compared to the emissions associated with an individual building heating system calculated in the same way.

For a given set of energy flows, e.g., electricity and gas imported to an energy centre, the CO_2 emissions are calculated using emission factors (kg kW^{-1} h^{-1}) for the fuels and electricity used or generated. When comparing DH/C emissions with individual building systems it is clearly important that the basis and underlying assumptions of the emissions factors used are the same.

Although a rigorous calculation of CO_2 emissions is a fundamental requirement if decisions are to be made on how best to deliver a low carbon future there is considerable difficulty in determining the appropriate emission factors to use as this section will show.

In section 6.4, heat pumps and CHP were compared using an electricity emission factor derived from the national energy statistics of 0.46 kg kW^{-1} h^{-1} which assumes gas-fired CCGT plant is generating electricity for the heat pump or being displaced by CHP. There is a need to evaluate CO_2 emissions over the lifetime of the plant when comparing heat supply options and as the grid is decarbonised in the future the emission factor to be applied will change.

A number of sets of emission factors are published:

- A comprehensive set of emission factors to be used in reporting of environmental impacts by companies is provided by Defra and BEIS [3, 4]. These are updated annually.
- BEIS have issued a set of emission factors to be used in evaluation of policies in Data Table 1 and 2a as part of the Supplementary Guidance to the HM Treasury Green Book [5, 6]. These include projections to 2050.
- The Standard Assessment Procedure SAP 10.2, which is used in the Building Regulations for compliance and for Energy Performance Certificates, has a set of emission factors which is based on an average of a 5-year projection from 2020 to 2025 [7, 8].

In comparing these sources and emission factors three issues arise:

- The scope of the emissions covered in the emission factor
- Whether the factors are average factors for the fuel and electricity or marginal figures
- The time-varying nature of emission factors and how averaging is applied

Scope

The definition of the emission factors published varies depending on the assumptions made. For example, for natural gas there are the following published emission factors:

SAP 10.2–0.21 kg kW^{-1} h^{-1}
BEIS Data Table 2a—0.183 kg kW^{-1} h^{-1}

The difference is accounted for by whether it is only emissions associated with the combustion of fuel on-site (referred to as Scope 1 in [3]) or whether upstream emissions are also included (Scope 3). The higher figure in SAP includes some upstream emissions associated with transport of the fuel and also includes an allowance for leakage as methane is also a greenhouse gas [8]. The term CO_2 equivalent is used to enable an overall GHG impact to be assessed including methane leakage. The SAP figure is based on the emission factors published for company reporting [4] which includes figures for natural gas and imported liquefied natural gas (LNG) as in table 9.1. As the proportion of LNG increases the emissions factor for mains gas will increase.

It is clearly important to be consistent in the definitions used. For example, it would be incorrect to use an emission factor for electricity derived from the Scope 1 fuel emission factor as in Data Table 2a [6] in the same calculation as emissions from boilers or CHP using a fuel emission factor that includes Scope 3 emissions, e.g., from SAP 10.2 [7].

Table 9.1. Emission factors for mains gas taken from [4].

	Scope 1	Scope 3	Total
	kg kWh^{-1} GCV	kg kWh^{-1} GCV	kg kWh^{-1} GCV
Natural gas	0.1825	0.0311	0.2136
LNG	0.184	0.063 67	0.2477
Weighted average (80% NG, 20% LNG)			0.2204

Average or marginal

UK government

The marginal emissions factor is recommended to be used by Government in the Supplementary Guidance to the HM Treasury Green Book [5] where Clause 3.22 states:

> *For estimating changes in emissions from changes in grid electricity use, analysts should use the (long run) marginal grid electricity emissions factors in Data Table 1*

And Clause 3.23 states:

> *There are complex mechanisms that determine the effects of sustained but marginal changes to the grid electricity supply (from either displacement with other generation or a demand reduction). A small reduction in grid electricity consumption will be met through a reduction in supply from a small subset of plants, rather than through an equal drop across all generation plants. Very temporary changes in consumption will likely only result in short run changes to generation levels, rather than changes in capacity. However, sustained changes in consumption will result in changes*

to generation capacity—in terms of the timing, type, and amount of generation plant built and/or retired—as well as changes in generation levels. Modelling undertaken by BEIS has estimated these longer-term dynamics, and they are reflected in the marginal emissions factors.

Data Table 1 provides the long-term electricity marginal emissions factor, assuming a constant change in demand over the year, for each year and assuming a scope 1 gas emissions factor of 0.183 kg kW^{-1} h^{-1}. This was set at 0.382 kg kW^{-1} h^{-1} for commercial electricity supply in year 2010 (equivalent to a CCGT efficiency of electricity supply of 48%) and from dynamic dispatch modelling a figure of 0.098 kg kW^{-1} h^{-1} was obtained for 2030. Interpolation was used for intermediate years so that for 2023 a figure of 0.244 kg kW^{-1} h^{-1} is given. This is much lower than the current operating marginal figure of 0.46 kg kW^{-1} h^{-1} calculated in section 6.4. This is partly due to these figures not including scope 3 emissions and partly due to the figures including the potential impact on new generation capacity and not simply 'operating marginal' (see below) emissions, see pages 5–6 in the Background Documentation report [6]. This report also states:

There remains much uncertainty, including for example, in the pace of innovation in the market, demand levels, the technical feasibility of some technologies, and the investment decisions of electricity generators.

UNFCCC

The UNFCCC methodology for calculating an emission factor for an electricity system [9] is used for analysing the impacts of projects supported under the Clean Development Mechanism (CDM). This similarly identifies two marginal impacts: operating marginal and build marginal, i.e., the impact of a reduction in demand may impact both existing power stations by displacing fuel used and also displacing the new capacity that might be built. For example, consider a grid dominated by coal-fired power stations. Adding a gas-fired power station would displace the coal-firing but doing so may prevent the construction of say a nuclear station with very low emissions, so the actual reduction in emissions is less than would be calculated solely using the operating margin approach.

Operating margin is defined as:

The generation weighted average emission factor for the power plants operating excluding low-cost/must run generators such as nuclear, wind, solar and low-cost biomass. In the GB case this remaining generation is dominated by gas-fired CCGT plant.

Build margin is defined as:

The generation weighted average emission factor for the power plants constructed in the last 10 years that supply at least 20% of the demand, based on their operation in the most recent year. In the GB case, very little capacity has been added as CCGT plant—the most recent CCGT plant coming on stream is Keadby 2 at 893 MW [10]. In contrast, new generation capacity commissioned since 2011 from [11] is:

On-shore wind (transmission connected): 3798 MW
Off-shore wind (transmission connected): 9209 MW
On-shore wind (distribution connected): 4915 MW
Solar (distribution connected): 12 628 MW

Hence it is clear that the weighted average emission factor for the build marginal will be close to zero. These renewable sources supply more than 20% of annual demand.

The emission factors from each of these are then combined as a weighted average to give the combined margin. Where the new generation in the build margin is wind and solar and not likely to contribute to peak capacity then the default weighting in the methodology is 0.75 operating margin and 0.25 build margin.

Using this methodology, and the emission factor for CCGT of 0.46 kg kWh^{-1} as calculated in section 6.4.3, the combined marginal figure would be:

$$0.46 \times 0.75 + 0 \times 0.25 = 0.35 \text{ kg kWh}^{-1}$$

SAP 10.2

The SAP electricity emission factors are based on average emission factors as SAP is used as a tool for reporting of environmental footprint in energy performance certificates. The use of average emission factors when used to compare energy options for new buildings, which is clearly a marginal activity, appears to be in contradiction to the guidance quoted above from [5]. It was found that CHP plants (and DH systems supplied by CHP) would be severely disadvantaged if the average electricity emission factor given as 0.136 kg kW^{-1} h^{-1} was used. So, for CHP plants, higher figures are used which are intended to represent the marginal electricity that would bc displaced by CIIP, see table 9.2.

As the flexible operation assumes that gas-fired CCGT is being displaced then taking the gas emission factor of 0.21 kg kW^{-1} h^{-1} it can be seen that an efficiency of grid electricity of 50% has been assumed. This is higher than the 45% efficiency calculated in section 6.4 from the national statistics.

Table 9.2. Emission factors for electricity generated by CHP from SAP 10.2 [7].

Type of CHP	Emission factor applied to electricity generated kg kW^{-1} h^{-1}	Definition
Export only	0.394	If all electricity is exported then operation will respond to the market price signals
Flexible operation	0.42	This requires plant to stop generating if gas-fired power stations are not operating on the grid
Standard	0.311	This operation assumes the CHP meets some on-site power use and so is likely to be less flexibly operated for economic reasons

Summary of heat emission factors under various assumptions

The calculations of CO_2 content of heat for boilers, CHP and heat pumps under the various emission factor assumptions described above are shown in table 9.3. It can be seen that if an operating marginal emission factor is used at 0.46 kg kW^{-1} h^{-1} CHP has a lower CO_2 content than a heat pump (assuming typical efficiencies relevant to DH). Using the BEIS figure CHP has the same CO_2 emissions as a boiler. For the UNFCC figure, both CHP and heat pumps save CO_2, but the heat pump has the lower CO_2 emissions. For SAP 10.2 the CHP has a CO_2 content about half that of boilers. If the same factor is applied to heat pumps then CHP has a lower CO_2 content than the heat pump but if the SAP 10.2 average electricity factor is applied (as in the SAP tool) then the heat pump has a very low CO_2 content.

Table 9.3. Impact of emission factors on CO_2 content of heat.

	CCGT marginal (as section 6.4)	BEIS long-term marginal	UNFCC marginal	SAP 10.2
Gas emission factor (kg kW^{-1} h^{-1})	0.21	0.183	0.21	0.21
Electricity emission factors (kg kW^{-1} h^{-1})	0.46	0.244	0.36	0.42 (CHP) 0.136 (HP)
Boiler heat emission factor (85% efficiency) (kg kW^{-1} h^{-1})	0.25	0.215	0.25	0.25
CHP heat emission factor (38% elec eff, 42% heat eff) (kg kW^{-1} h^{-1})	0.084	0.215	0.174	0.12
Heat pump heat emission factor (CoP = 3) (kg kW^{-1} h^{-1})	0.153	0.081	0.12	0.14 (CHP) 0.045 (HP)

Limitations of the SAP 10.2 approach

When comparing individual systems to DH the same emissions factor should be used for both systems. The marginal electricity emission factor should apply to small changes in demand, whether increases or decreases, regardless of technology used. This is not the case when applying the SAP 10.2 emissions factors which use an average electricity emissions factor for say a heat pump in an individual building (and also in a DH Energy Centre) and a marginal figure for the CHP plant. This means that SAP will always result in a lower emission rate for an individual heat pump system when compared to a CHP-based DH system even though this is unlikely to represent the actual CO_2 emissions that will occur and potentially puts existing CHP-based DH at a disadvantage. The use of an average emissions factor implies that when a heat pump is installed, and electricity demand increases, this additional demand is met by all generators on the system weighted in relation to their annual output. Given that the heat pump's operation will be mostly in the winter it is unlikely that, e.g., solar output would increase in winter to supply the increased demand. Nor would nuclear output change as it already runs at maximum output and new capacity is being added slowly, largely to replace existing plant that is being retired rather than to meet increases in demand.

Green Heat Networks Fund

The Green Heat Networks Fund (GHNF) requires calculations of CO_2 based on the electricity emission factor from Data Table 1 (long run marginal for commercial) [6] and the gas scope 1 gas figure used for company reporting [4]. The electricity emission factor used has been developed assuming constant changes in demand over the year which is not the case for DH with its variable demand weighted towards the winter and so is likely to underestimate the benefits of CHP and overestimate the benefits of heat pumps.

Time-varying emission factors for electricity

The difference in the figures for CHP in table 9.2, depending on whether its output is fixed, i.e., running at a constant rate through the year or operating flexibly is a recognition of the fact that the marginal plant on the grid will vary depending on the time of day, and the output of variable renewable energy sources. It can be seen that the use of a single emission factor, however derived, has limitations in accurately representing the actual CO_2 savings.

The mix of electricity sources supplying the grid is continuously changing and so the average emissions factor also varies throughout the year. This is available from the Electricity System Operator (ESO) (https://www.carbonintensity.org.uk/) and from Gridwatch UK (https://gridwatch.co.uk/co2-Emissions). The ESO website includes a 4-day lookahead forecast as well as regional information. SAP 10.2 recognises this variation and has published average electricity emission factors for each month of the year, with higher figures in the winter period, see table 12d in [7]. However, all these sources provide average emission factors not marginal emission factors.

The operating marginal emission factor will vary on an hour-by-hour basis through the year as the mix of plant at the margin varies. In recent years, the marginal plant has always been gas-fired CCGT apart from a small number of hours where coal-fired plant is needed to operate to meet the demand. In the future there will be times when the marginal plant, i.e., that plant which will modify its output in response to changes in demand, will be wind or solar, initially at times of high wind or solar output and low national electricity demand. Section 10.2 shows how future scenarios might develop and how the marginal plant on the system might vary through the year as a result.

A DH production plant supplying a heat demand profile which varies over the year will not operate uniformly throughout the year and different types of plant will operate in different patterns making use of thermal storage to minimise operating costs. The BEIS guidance [5] notes that the marginal figures provided assume a uniform change in electricity demand through the year, which is clearly not the case for a DH or DC scheme, and recommends that a bespoke analysis is carried out for other operating patterns.

Therefore, rather than a single emissions factor for electricity as is commonly used in options appraisals, a time-varying emission factor including future projections for each year of the analysis period is needed. This in turn will influence the plant to be selected for DH, how it is operated and how thermal storage is exploited and potentially lead to lower CO_2 emissions as a result of the more accurate analysis.

The uncertainties in making future projections are significant. The marginal emission factors will depend both on changes in demand, e.g., the uptake of electric vehicles and heat pumps and the rate at which new low carbon generators: wind, solar, nuclear and gas-CCUS are constructed. Strategies which take account of these uncertainties are therefore of interest, including DH systems supplied from a mix of technologies (e.g., both CHP and heat pumps) which can be operated flexibly with thermal storage and adapt to changing circumstances over time. This is discussed further in section 10.2.

9.2.2 Other greenhouse gases

Mention has been made already of the need to account for other greenhouse gases when calculating emission factors, such as methane leakage from natural gas supplies.

For both DH and DC systems there is a significant potential impact from the leakage of refrigerant gases—for DH systems based on heat pumps and for chillers in DC systems. The impact of refrigerants on global warming is given by their global warming index which is related to CO_2 equivalence—for commonly used refrigerants these are given in table 6.1 in section 6.3 and typical leakage rates are given in CIBSE TM65 [12].

For heat pumps, an individual building solution may use centralised heat pumps or a VRF system typically using R32 or R134a whereas for the DH heat pump, a low or zero global warming potential (GWP) refrigerant may be used such as ammonia. This can create a differential in emissions in favour of DH.

For DC a similar situation occurs. It is more likely that a low or zero GWP refrigerant would be used in a central plant than in chillers in individual buildings partly due to safety considerations and availability of cost-effective plant. Again this will create a differential in refrigerant emissions in favour of DC.

9.2.3 Comparison of greenhouse gas emissions for individual buildings and DH systems

When comparing DH with gas boilers in individual buildings there will clearly be a significant advantage if the DH system is using CHP or heat pumps.

When comparing DH with individual building heat pumps the position will depend on a number of factors and a detailed analysis is required. Table 9.4 summarises the main climate change impacts that need to be assessed in the comparison.

Table 9.4. Comparisons of DH with individual heating systems—climate change impacts.

Factor of influence	Individual	DH	Balance in favour of
Delivery temperature	To suit building, can be lower than 60 °C for space heating	Probably more than 60 °C for legionella protection; temperature set by the building with highest temperature needs on the network	Individual
Source temperature of heat pump	Likely to be air	Could be from elevated temperature, e.g., data centre or river	DH but case specific
CoP of heat pump	Limited by technology and scale	Could be higher due to advanced technology and scale	DH but probably small benefit
Hybrid solutions including CHP and peak boilers	Unlikely to be used as would be more complex and expensive	CHP and heat pump solution may have lower emissions, peak boilers would increase emissions	DH but depends on mix and assumptions on emission factors for electricity
Network losses	None or negligible	Could add 5%–20% to emissions	Individual
Network pumping and ancillary electricity use	None or negligible	DH pumps and electricity use in energy centre ca. 2% of heat delivered	Individual
Refrigerant leakage	Higher GHG refrigerants likely to be used, may not be monitored carefully	Could use low or zero GHG refrigerants. More sophisticated monitoring and control of leakage	DH
Operational flexibility	Limited as thermal storage often impractical or expensive	Thermal storage and hybrid plant allows operation to minimise emissions	DH
Alternative heat sources	Unlikely to be available	Can exploit sources such as industrial heat recovery, heat from power stations, deep geothermal, solar thermal with interseasonal store	DH but case specific

9.2.4 Other environmental impacts

A full comparison of environmental impacts between individual building systems and DH needs to consider other issues.

Air quality
Where DH and individual systems are both supplied from heat pumps then there would be no difference in air quality impacts. However, if the DH system includes gas boilers or CHP then an analysis of NOx levels as a result of the addition of the DH energy centre needs to be undertaken and compared with the business as usual case of retaining gas boilers in buildings or with heat pumps in each building. Where the DH is supplied from remote power stations including energy from waste plant then the addition of DH to these sources will not alter their existing emissions and air quality would be unchanged.

Water use and impact on water bodies
DC systems may use evaporative cooling towers which will involve higher water use than might be used by air-cooled chillers in individual buildings. Where river water heat pumps are used the water will be returned at a slightly lower temperature which is not normally of concern. However, high extraction flow rates and/or discharge rates may impact local wildlife or risk erosion of the river banks or river bed and so these concerns will need to be addressed.

Visual impact and noise
Another potential disadvantage of a DH scheme is likely to be the visual impact of a new energy centre which needs to be carefully considered and a suitable architectural approach developed (see section 6.15). As part of the design, acoustic assessments will be needed.

Air source heat pumps (ASHP) in individual buildings and dwellings will also have a visual impact and also a noise impact especially where there is the cumulative effect of every house in a street or every flat in a block having an ASHP.

Embodied energy
There is increasing recognition that just considering operational emissions does not present a full picture of the impact on the environment, see CIBSE TM65 [12]. An example is the recent debate about whether to demolish and rebuild or to refurbish existing buildings.

For DH a legitimate question is how the embodied energy associated with the construction of the energy centre, heat network and interfaces compares with that of individual building systems. A detailed treatment of this issue is outside the scope of this book, however, it is clear that the DH network will involve greater use of materials and construction energy than individual heat pumps and this should be taken into account when comparing energy systems. Such an analysis also needs to consider the lifetime of the components. It would be expected that individual heat pumps will need to be replaced several times within the lifetime of a heat network.

9.3 Economic

An economic analysis will aim to compare the total cost of heat/coolth supply for the DH/C option with one or more alternative methods of heating/cooling provision.

This may be the continuation of the existing heating system, e.g., gas boilers, (business as usual) or its replacement with a low carbon alternative, e.g., a building scale heat pump (the counterfactual).

A DH project is likely to offer an economic advantage compared to individual building solutions in the following principal ways:

- Economies of scale for the low carbon plant, for both capital and maintenance.
- Diversity of demand so the central plant capacity will be smaller than the sum of individual building plant capacities.
- The use of peak boilers meaning that the high cost, low carbon plant need not be sized to meet peak demands.
- The use of thermal storage to enable operation at times to minimise energy costs.
- A single electricity supply to an energy centre instead of multiple upgrades that might be needed to each building.
- Simpler modifications to the existing heating systems and less impact on space in the building.
- The life of the network and major plant items will be longer than the smaller scale equipment used in buildings so replacement costs are lower.

Against these benefits needs to be set the capital cost of the DH network and its operating cost.

As it will be necessary to compare both capital and operating costs for each option these are best presented as the net present cost for the supply of energy. To do this an analysis period needs to be assumed together with a discount rate which enables operating costs in each year to be capitalised. These together with any replacement costs and initial capital costs will summate to the net present cost. It should be noted that this approach does not involve setting a heat selling price, it is simply comparing lifetime costs to meet the energy demand.

In assembling the costs for the comparison care is needed to include all appropriate costs and to exclude as far as possible taxes, grants and subsidies. The aim is to analyse the options from the view of the national economy rather than for the specific project which is dealt with in the commercial case.

In many projects, each option will have different lifetime CO_2 emissions and plotting the results on a graph of NPC (y-axis) and CO_2 (x-axis) can be helpful. As an example, figure 9.2 shows four options. The BAU point is a low-cost option as there is no capital expenditure but has high CO_2 emissions, e.g., a building with existing gas boilers. Option A is the opposite—it has very low emissions but there is a high cost, probably for new investment in a low carbon system within the building. Options B and C are intermediate options which although having a higher cost are moving towards the graph's ideal point at the origin of aiming to achieve low-cost and low emissions and might represent two different DH options. In considering whether C is better than B it is useful to calculate the marginal cost of CO_2 reduction

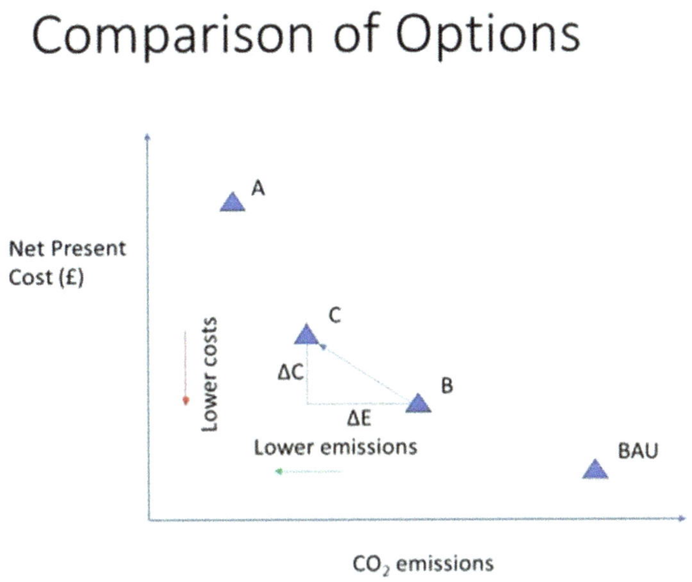

Figure 9.2. Assessing economic and environmental benefits for a range of options.

in £/tonne which can be easily seen and calculated as $\Delta C/\Delta E$, see figure 9.2. This can then be compared with a benchmark cost of CO_2 emitted as given in [6].

9.4 Security of supply and safety

9.4.1 Strategic security of supply

In the past, DH has contributed to the security of a country's energy supply by reducing fossil fuel consumption through CHP and thus extending the life of the nation's reserves and reducing the need for imports. This has been particularly the case for Scandinavian countries which were largely reliant on imported energy. In the UK, DH will contribute to improving security of supply both by continuing to use gas more efficiently with CHP and transitioning to electric heat pumps which will largely use power from indigenous resources such as wind, solar and nuclear.

9.4.2 Security of supply for customers

Customers will rightly demand a high level of security of supply from DH as for any other heating system. Whereas failures in individual systems occur regularly only one customer is affected. A failure in a DH system would affect many customers and even if the total failure time is the same as multiple individual failures the impact on the reputation of the DH company would be significant. As a result, considerable effort is devoted to providing a very reliable supply. Typically, the following are included in a DH system:

- Multiple heat sources including standby boiler capacity so that a failure of any one plant item would still enable peak demands to be met (see section 6.11).
- Multiple distribution pumps so that a single pump failure would not impact the supply.
- Power supplies which include back-up supplies or standby generators so that a total loss of electricity is very unlikely.
- An uninterruptible power supply (UPS) for the control systems so that any short-term interruptions in power supply do not result in loss of information and delay and disruption in restarting the plant.

This still leaves some potential single points of failure:

- Gas supply for schemes dependent on gas for all of the heat sources.
- The heat network itself unless this is designed with loops which create some redundancy.

The reliability of the gas network is very high and for safety reasons any failures would be dealt with quickly.

The heat network itself is also very reliable provided a high standard of construction and quality control has been achieved (see section 5.8).

There may be failures in the interface equipment and HIUs however these are likely to be minor and easily rectified provided suitable parts are kept on-site or available at short notice and would only affect one customer.

Aside from the reputational damage any failure will lead to loss of revenue from heat sales and potentially penalty payments to customers so there is a strong incentive on DH companies to achieve high reliability. It is important that there is an incentive on the construction company to achieve a high quality build either through the use of a design, build, operate form of contract or by means of extended warranties on the installation.

The biggest risk to a DH network is probably third party damage to the installation as a result of adjacent works to other utilities or the highway. The provision of accurate information on the location of the DH pipes in line and level to other utility providers is therefore critical, as required by Clause 79 of the New Roads and Street Works Act [13].

9.4.3 Health and safety

A DH supply is inherently safer than other heating options.

With natural gas and LPG there is the risk of explosions which typically result in 31 incidents a year (average of last four years data). Poorly maintained combustion appliances can result in carbon monoxide poisoning with 90 incidents and 3000 dangerous appliances reported a year (average of last four years data), see HSE statistics [14].

Other fuels stored on-site, (oil, biomass) and electric fires and other highly rated electrical equipment can represent a fire risk.

The use of wood-burners and gas-fires in dwellings can result in poor air quality.

In contrast, the relatively low pressures and temperatures used for the water circuits in modern DH systems mean that these safety concerns are not present. The higher risk items of plant are taken out of buildings and placed in an energy centre where risks can be well controlled. For vulnerable households low surface temperature radiators can be used and exposed pipework boxed-in so that even scalding risk can be minimised.

9.5 The nature of the market and customer choice and experience

9.5.1 Residential customers

A report by the ADE [15] identified a number of barriers to the uptake of low carbon and energy efficient technologies by the householder and stated:

> *Many householders find it difficult to understand the choices on offer, which of those would best suit their particular circumstances, and how their choices relate to the wider energy system*

and

> *Policies, markets, and people—even the well-meaning and well-informed—intermix to create a web of confusing messages that leave people feeling unable to act.*

Advantages for the householder

For the householder, DH offers a number of advantages compared to installing individual heat pumps.

Technical risks and quality

The DH installation will be carried out by the DH company who will have a strong long-term interest in delivering a high quality installation which will lead to lower operation and maintenance costs. This is in contrast to individual heat pumps where the installation company has no ongoing interest. The householder is more likely to have confidence in the quality of the installation and avoid risks of poor design, specification and equipment. The technical risk on the equipment is normally taken by individuals when purchasing heat pumps whereas for DH the risk lies with the DH company.

The DH industry in the UK has seen significant improvement in quality over the last few decades. It has benefited from the development of European standards for key products such as pre-insulated pipes and heat meters. A number of industry guides have been produced including CIBSE CP1.2 [16]. There is now a system of testing of HIUs which is well established and has been extended to cover additional types of HIU (see section 4.3.2). This has already resulted in advances in performance and quality. The latest development is the Heat Network Technical Assurance Scheme which will provide certification of quality at both the design and construction stages (see https://fairheat.com/developing-a-heat-networks-technical-assurance-scheme-in-the-uk/).

Transaction costs
With an individual heat pump the householder will have to organise the work including finding suitable installers, obtaining quotations and comparing these. There may also be a need to apply for planning permission, or for permission to connect a heat pump to the electricity supply. With DH, it will be the DH company that will be responsible for these transaction costs, rather than the householder which could be a major factor in choosing DH as the low carbon solution.

Finance
With current business models for DH it is unlikely that householders will need to pay an initial connection charge as the costs of the DH system will be financed centrally with repayments made from the revenue from heat sales. The barrier of having to raise capital for an individual building heat pump system is therefore avoided. In addition, the DH system will be financed at lower interest rates than available for householders and over a longer time period. This will lead to lower costs of financing which will be reflected in the heat selling price.

Standards of service and operating temperatures
DH will typically be designed to provide higher temperatures than would be selected for individual heat pumps. DH heat pump systems with a wider range of refrigerants and heat sources such as energy from waste plants together with the use of peak boilers could supply temperatures in the region of 85 °C–90 °C at peak thus avoiding the need to change radiators, although rebalancing to achieve low return temperatures will be necessary (see section 7.2). This is reflected in the economic benefits of DH but also reduces the disruption to the customer's premises.

Also significant for dwellings is the ability of DH to deliver a high flow of heat energy sufficient to provide instantaneous hot water, typically 35–50 kW. This rate of heating cannot be delivered by a heat pump that typically would be 5–10 kW. As a result, individual heat pump systems need to have a hot water cylinder to be charged up by the heat pump. This has a number of implications for the householder:

- Combi boilers are the most popular choice for gas systems; the market share in England was 60% in 2019 [17]. Thus most homes will not have an existing cylinder and there will be a need to find suitable space for the new cylinder. In many homes this will be a serious drawback.
- The householder with a combi boiler will be used to having as much hot water as needed any time of day or night without the risk of draining the tank and having to wait to reheat it.
- Combi boilers normally use mains pressure hot water and to provide the same service pressure a new cylinder will need to be fitted with the necessary safety devices and expansion vessels increasing the cost.

In contrast, a DH system can provide the equivalent of a combi boiler in terms of hot water service, can use mains pressure and the HIU would occupy a space no more than a typical gas combi boiler.

The above drawbacks of the individual building heat pump system apply equally for the ambient temperature network system described in section 10.8.

An individual air source heat pump system will need to be located outside the dwelling and this could also be difficult to site without detracting from a garden amenity area. There needs to be space around the unit for free air flow. It is possible that in some locations (say on balconies in blocks of flats) there could be a nuisance from noise either to the owner or a neighbour. In contrast, a DH system would have no equipment installed externally and no risk of noise.

In general, it can be said that DH can be installed to supply any type of building and is proven technology. Heat pump installations are likely to prove difficult to install in some buildings due to space, noise, complexity and the need to modify existing heating and hot water systems.

Potential drawbacks for the householder

Lack of choice of supplier and mandating of connection

It is self-evident that DH will be most cost-effective if all buildings in a given area are connected to the heat network as costs are shared over the maximum number of customers. The proposal for cities to be zoned to identify where DH is most cost-effective and where connection would be mandated could mean that householders would not be able to choose whether to connect or not (zoning is discussed further in section 10.9). If the zoning methodology is accurate it should mean that anyone mandated to connect to DH would pay less for heating than if they installed their own low carbon system. This reassurance should mean that the lack of choice is not seen as a major drawback.

Lack of statutory consumer protection

At present, customers on heat networks do not benefit from the regulatory protection given to electricity and gas customers by the regulator Ofgem. Instead, there has been a voluntary scheme set up by the Heat Trust and supported by the industry to set minimum customer service standards which are intended to broadly mirror the regulated energy supplies. These standards are contained in the scheme rules of the Heat Trust [18]. Members of the Heat Trust align their heat supply contracts to meet the standards set out in the scheme rules and pay a fee per property to the Heat Trust to cover their administration costs. One important provision is that the Heat Trust provides a route to the Energy Ombudsman to deal with complaints that have not been resolved.

In 2018, the Competition and Markets Authority prepared a report on DH [19] which recommended putting customer protection for DH on a statutory basis. This has now been taken forward with proposed measures to be included in the Energy Bill, 2023 [20] including for Ofgem to have a responsibility for regulating heat supply contracts from DH companies.

Need to contract with another utility

A DH connection will involve the customer dealing with a new utility with meter reading and paying bills. A separate contract would be set up for the provision of heat. This is in contrast with an individual heat pump system where the existing electricity contract will remain in place. However, typically the DH supply contract will mean the gas supply contract is no longer required. For the ambient temperature network system the energy charges would be with the electricity contract however there would be another contract for the supply of the ambient water which would be charged on a fixed price basis rather than a metered supply.

A summary of the impact on householders is given in table 9.5.

Table 9.5. Impact on householders of connecting to DH.

Feature	Individual heat pump	DH	Favourable to DH?
Quality of installation	Contractor installs heat pump but no long-term interest	DH company provides install service and has a long-term interest	yes
Transaction cost	High	Low	yes
Radiators	Need replacing	Can be kept and rebalanced	yes
Hot water service	Requires cylinder, space and cost	Instantaneous compact	yes
External space	Needed for ASHP unit	Not required	yes
Choice	May be a ban on new gas boilers in the future	Connection may be mandated if in a heat network zone	Both options comparable in the future
Consumer protection	Price caps and supply standards regulated by Ofgem	Awaiting regulations to be implemented	Both options comparable in the future
New utility	Not required	Required but gas contract would be dropped	Both options comparable in the future

9.5.2 Commercial and institutional customers

In the past, these customers are likely to have been attracted to DH mainly by lower cost heat supply compared to continuing with their current heat supply, provided they have been satisfied that the reliability will be similar to their current system. If there was also an environmental benefit that would be welcomed but the decision would be based on their commercial position. In the future, companies and institutions are increasingly likely to take account of their environmental responsibilities and then DH will be compared with the alternative low carbon solution that could be delivered at a building or site level. In this comparison the DH option will not necessarily be the lowest CO_2 but may well offer the least cost and least disruptive solution, especially where the customer has limited access to capital for a major change to the heating or cooling systems. Where the alternative proposed is for an on-site ASHP system the level of disruption could be significant as space will be needed for the heat pump, the power supplies may need upgrading and the heating system itself may need replacing or modifying. In contrast, the DH offer would involve a reduction in space (if existing boilers are removed), no change in power supplies and only minor modifications to the heating system.

The sleeving arrangement allowed for in the Green Heat Networks Fund grant scheme could be useful in allowing new investments in low carbon heat production to be allocated to a new customer so that the DH offer is better able to compete with an on-site solution.

In assessing the benefits of DH over an alternative it is normal to assess the site's costs in terms of:

- Fuel and electricity costs
- Fixed and variable maintenance costs
- Avoided capital replacement costs or avoided new low carbon technology investment

In addition, there are associated management costs which are difficult to quantify to cover for the procurement of energy, maintenance work and managing capital projects. For many organisations, management time needs to be devoted to the core business not to a complex change to a low carbon heating or cooling system. These management responsibilities are particularly acute for DC where maintenance of cooling towers is critical to avoid legionella. DC is often sold as a premium service which means that building operators are often willing to pay more to avoid the responsibility and costs of maintaining their cooling plant, effectively outsourcing this part of their operations, in other words 'cooling as a service' [21].

9.6 DH/C projects in the five case model

Many DH/C projects in the UK have been financed in part by central or local Government. The preferred approach by Government is to prepare a business case using the five case model. Each of these cases are briefly discussed below with the main points relating to DH/C systems highlighted.

9.6.1 Strategic case

The strategic case for DH/C is normally very strong as it is recognised as an effective way to address climate change which many local authorities have declared an 'emergency'. Other strategic issues that may be addressed through DH/C include: dealing with the need to replace ageing boilers, improving local air quality, retaining more money in the local economy and creating additional jobs in the community. Without DH/C the likelihood is that gas boilers will continue in use. The alternative strategy of replacing gas boilers with individual building heat pumps will need to be evaluated.

9.6.2 Economic case

The economic case will be presented in terms of comparisons on an NPC basis as outlined in section 9.3 for a range of options for the provision of future heating and cooling to buildings. The main options to be evaluated for heating are likely to be: business as usual or 'do minimum', replace boilers with DH, replace boilers with individual heat pumps, or replace boilers with individual heat pumps using an ambient temperature network as the heat source. The DH option may have sub-options with different heat sources being evaluated. These supply options may be evaluated with and without investments in reducing demand.

The economic case needs to exclude taxes, grants and subsidies as far as possible so as to demonstrate that there is an underlying economic benefit for society. The value of the CO_2 reduction may be included as an income stream using the social cost of carbon. A discount rate of 3.5% in real terms would be used and the case for DH would be made if the NPC for the preferred DH option has the lowest NPC.

When comparing DH for a large city area with individual building heat pumps or hydrogen it is worth noting that there may be costs for upgrading of the electricity and gas distribution networks if these options are to be pursued across the area, even if it is difficult to assess these costs at this stage.

9.6.3 Commercial case

In contrast to the economic case the analysis will present the commercial deal from the perspective of an investor. This will involve the setting of energy sales prices (heat, cooling and electricity where CHP is used), estimates for capex and energy imports. The analysis will include taxes and subsidies where these can be reliably predicted. Typically, the key parameter is the internal rate of return (IRR). A range of sensitivities will be used to see how the IRR changes in adverse or positive scenarios, especially around future energy price changes.

The commercial case will set out how the contracts will be structured and how the main risks will be allocated between the parties. A general principle is that risks should be allocated to the party best placed to manage the risk. A number of arrangements have been used successfully in the UK, e.g.:

(A) Concession model

In this model the local authority (LA) would invite bids to develop a DH/C business in a defined area and once awarded the LA would have limited involvement, acting as one of the customers, the planning authority and monitoring the performance of the concessionaire against the terms of the concession. These terms may include a measure of price control to protect customers. UK examples are: Queen Elizabeth Olympic Park, Citigen in City of London. In general, risks are transferred fully to the concessionaire.

(B) Design, build, operate and maintain (DBOM)

Here the LA will seek bids for a contractor to design, build, operate and maintain the DH/C system but ownership is retained by the LA. Heat may be sold directly to customers or sold to the LA for onward sale. Heat sale prices are normally fixed and adjusted annually via an indexing mechanism. Most construction and operational risks are held by the contractor, however, volume of heat sales and price risk are with the LA. The latter can be mitigated to some extent through the heat supply contracts. The LA would act as supplier of last resort.

(C) Separate design and build (D and B) and operate and maintain (O and M) contracts

This arrangement provides the LA with more flexibility and more direct involvement. The O and M contract could be let for a shorter duration and to O and M specialists. Having a separate D and B contract may widen the pool of construction contractors. Although risks are transferred in a similar way to B) there are risks associated with the interface between the two contractors to be managed. The O and M contract may require the contractor to sell heat to the LA or to simply charge for operating and maintaining the plant with the input energy costs paid by the LA who also sells heat to the customers. This avoids the need to set up an indexing arrangement with the contractor in relation to energy, but suitable plant availability and efficiency incentives and penalties are needed (e.g., through Key Performance Indicators).

(D) Independent ESCo and community interest companies

As an alternative to the above, a private sector ESCo may develop a scheme or a community interest company may be set up. The latter may be most appropriate for smaller schemes with strong customer support from the start of the project. A further variant is a joint venture arrangement between the LA and a private sector company.

More detail on the various business structures that can be used are described under Objective 2.13 in the Code of Practice CP1.2 [16].

9.6.4 Financial case

This case will demonstrate how the necessary funds needed for the capital investment will be raised. The funding streams may involve off balance sheet, borrowing or grants from central or regional Governments. The impact on the revenue account for each year of the project and how borrowing is repaid needs to be clearly established. Evidence of commitment by lenders, grant providers, customers and other stakeholders is needed to confirm that sufficient funds and income streams will be available when needed.

9.6.5 Management case

A DH/C project will have a number of stages and throughout there needs to be a clear management structure with suitable governance. The stages to consider would typically cover:

- Feasibility
- Design development
- Consenting

- Commercialisation (outline business case)
- Procurement
- Financial close (final business case)
- Construction
- Operation
- Further expansion

The management of change is critical as few projects proceed without changes from the original scope for good reasons (e.g., an additional customer joins the scheme). Active risk management is also advised to reduce the impact of adverse risks on the project outcomes regardless of where these risks are held.

9.6.6 Team structure

Typically, a LA will be advised by a team of professionals. The lead organisation for developing each of the five cases will differ and although all parties will need to work collaboratively, the lead organisation is normally as follows:

- Strategic case—the LA or other project sponsor
- Economic case—the technical consultant who normally develops a techno-economic model to evaluate options
- Commercial case—the lawyer or commercial adviser
- Financial case—the commercial adviser and the LA finance team
- Management case—the project manager and the LA contract management team

It is also possible that a private sector partner will have been selected by the time the final business case is developed and will be able to use their experience of delivering projects to advise on the development of these cases.

Bibliography

[1] World Energy *Trilemma Index 2022* World Energy Council (https://worldenergy.org/transition-toolkit/world-energy-trilemma-index)

[2] HM Treasury 2018 *Guide to Developing the Project Business Case* (https://gov.uk/government/publications/the-green-book-appraisal-and-evaluation-in-central-governent)

[3] HM Government March 2019 *Environmental Reporting Guidelines: Including Streamlined Energy and Carbon Reporting Guidance* (https://gov.uk/government/publications/environmental-reporting-guidelines-including-mandatory-greenhouse-gas-emissions-reporting-guidance)

[4] HM Government 2022 *GHG Conversion Factors (Full Set) Spreadsheet* (https://gov.uk/government/publications/greenhouse-gas-reporting-conversion-factors-2022)

[5] BEIS January 2023 *Valuation of Energy Use and Greenhouse Gas (GHG) Emissions Supplementary Guidance to the HM Treasury Green Book on Appraisal and Evaluation in Central Government* (https://gov.uk/government/publications/valuation-of-energy-use-and-greenhouse-gas-emissions-for-appraisal)

[6] BEIS January 2023 *Valuation of Energy Use and Greenhouse Gas (GHG) Emissions Background Documentation and Data* Tables *1–19* (https://gov.uk/government/publications/valuation-of-energy-use-and-greenhouse-gas-emissions-for-appraisal)

[7] BRE 2022 *The Standard Assessment Procedure (SAP10.2)* (https://bregroup.com/sap/sap10/
[8] BRE 2020 *SAP 10 Technical Paper S10TP-15 CO_2 and Primary Energy Factors for SAP*, version 1.1 (https://bregroup.com/sap/sap10/)
[9] UNFCCC 2018 *Tool07, Methodological Tool, Tool to Calculate the Emission Factor for an Electricity System* (https://cdm.unfccc.int/methodologies/PAmethodologies/tools)
[10] SSE Thermal 15th March 2023 *Europe's Most-Efficient Power Station Enters Commercial Operation in the Humber* (https://ssethermal.com/news-and-views/2023/03/europe-s-most-efficient-power-station-enters-commercial-operation-in-the-humber/)
[11] BEIS 2022 *Digest of UK Energy Statistics, 2021* (https://gov.uk/government/statistics/digest-of-uk-energy-statistics-dukes-2021)
[12] Chartered Institution of Building Services Engineers (CIBSE) 2021 *Technical Memorandum TM65, Embodied Carbon in Building Services: A Calculation Methodology* (CIBSE)
[13] HMG 1991 *New Roads and Street Works Act, 1991* HM Government (https://legislation.gov.uk/ukpga/1991/22/contents)
[14] HSE 2022 *RIDGAS: Gas-related Incidents Reported under RIDDOR in Great Britain* (https://view.officeapps.live.com/op/view.aspx?src=https%3A%2F%2Fwww.hse.gov.uk%2Fstatistics%2Ftables%2Fridgas.xlsx&wdOrigin=BROWSELINK, ridgas.xlsx)
[15] ADE 2020 *Laying the Foundations for Net Zero: Putting Households at the Heart of the Energy Transition* Association for Distributed Energy (https://theade.co.uk/resources/publications/laying-the-foundations-for-net-zero-putting-households-at-the-heart-of-the)
[16] CIBSE 2020 *Heat Networks: Code of Practice for the UK CP1.2* (CIBSE)
[17] Statista 2023 *Boiler Types in Dwellings in England 2019* (https://statista.com/statistics/292259/boiler-types-in-dwellings-in-england/)
[18] Heat Trust March 2023 *Scheme Rules* (https://heattrust.org/the-scheme-rules)
[19] Competition and Markets Authority 2018 *Heat Networks Market Study Final Report* (https://gov.uk/cma-cases/heat-networks-market-study)
[20] HMG March 2023 *Energy Bill Policy Statement, Heat Network Market Framework* (https://gov.uk/government/publications/energy-security-bill-policy-statements-and-draft-regulations)
[21] Clean Cooling Collaborative *Cooling as a Service* (https://cleancoolingcollaborative.org/wp-content/uploads/2018/07/Cooling-as-a-service-knowledge-brief.pdf Cooling-as-a-service-knowledge-brief.pdf) (cleancoolingcollaborative.org)

Chapter 10

Recent developments and future opportunities

10.1 Introduction

The first heat strategy was published by the Government in 2013 and subsequently the Heat Networks Delivery Unit was formed. Grants and loans for development work and capital projects have been provided through the Heat Networks Investment Project and more recently the Green Heat Networks Fund.

In March 2023, the UK Government published its energy strategy Powering up Britain [1], which brings together both net zero ambitions and energy security considerations, particularly in the light of the conflict in Ukraine.

In this document Heat Networks are described as 'vital to making net zero a reality' with £220 m being allocated and a commitment to continue providing grant support until 2028:

> *Heat networks are vital to making net zero a reality in the UK. In high density urban areas, they are often the lowest cost, low carbon heating option. This is because they offer a communal solution that can provide heat to a range of homes and businesses by capturing or generating heat locally. We are continuing to grow and decarbonise the UK heat network market through the Green Heat Network Fund and the Heat Network Efficiency Scheme. We can now confirm that capital support will be extended to 2028 to facilitate the continued growth of low carbon heat networks, including £220 million for the Heat Network Transformation Programme over 2025/6 and 2026/7.*

There is little information in the strategy on the wider role that heat networks could have, although other parts do provide indicators of the Government's direction. Heat network developers need to be aware of these scenarios to maximise their future opportunities such as:

doi:10.1088/978-0-7503-5286-4ch10

- Obtaining lower electricity costs through greater use of thermal energy storage in a future with more variable renewable generation.
- Hydrogen as a low carbon fuel in CHP plants.
- Carbon capture usage and storage (CCUS) with continued use of fossil fuels in power plants, potentially designed as CHP for DH.
- Small modular nuclear reactors designed as CHP for DH.

In addition to the broader opportunities presented under these headings there are other future opportunities specific to DH/C which are discussed in this chapter:

- District cooling (DC)
- Combining DH and DC including the use of ambient networks
- Zoning policies
- Reducing costs of DH with novel designs

10.2 Additional benefits from thermal storage with more renewable electricity generation

10.2.1 The nature of renewable electricity and the need for electricity storage

If the UK is to meet its climate change goals, there will need to be a significant expansion of low carbon electricity generation not just to displace current fossil fuel use but to meet increasing demand for electricity from electric vehicles and heat pumps. New capacity is likely to be dominated by wind and solar PV, especially offshore wind. Nuclear power is targeted to deliver 25% of our electricity demand by 2050 and a contribution from fossil fuel stations fitted with carbon capture use and storage (CCUS) is also proposed in the latest Government energy strategy [1].

The variability of wind and solar generation creates challenges in managing the electricity grid [2]. Wind and solar output is always maximised as operating costs are very low (except if there are transmission and distribution constraints). Nuclear stations are typically operated as base load for technical and economic reasons and so are not able to modulate output to assist in managing variable renewable generation. Biomass and energy from waste plants also have low net operating costs due to renewable subsidies or gate fees and also run at maximum output. At present, there is limited electricity storage available, mainly from the pumped hydro station at Dinorwig with 1.7 GW maximum output (https://www.fhc.co.uk/). The interconnectors to the continental grid and Ireland provide some flexibility by both exporting and importing power. However, it is the gas-fired CCGT stations that are used to supplement renewable and nuclear generation and it is their output that is adjusted to match the demand.

Electricity demand varies across the day, being lower at night and peaking between 1630 and 1830. It also varies weekly with lower demands at weekends and seasonally with more demand in winter as a result of greater use of lighting and heating. In the future, electricity demand is expected to increase from growth in electric vehicles (EVs) and greater use of both DH and building level heat pumps for space and water heating, the latter resulting in a relatively higher winter demand than at present. There will also be changes in demand due to: population growth, demographic changes, trends in industry, energy efficiency improvements and climate change.

To achieve a net zero energy system for the UK there will be a need to add more variable renewable generation. This will create a greater need for storage so that at times of surplus the generation need not be shut down and at times of shortfall the use of gas-fired peaking plant can be reduced.

The relevance for DH/C is that it is possible to store thermal energy efficiently not just within a 24 h period but over days, weeks or even interseasonally. DH/C supplied by heat pumps and with thermal energy storage (TES) could therefore have a role in managing electricity demand patterns over these longer periods—to use renewable electricity at times of surplus and to limit electricity use at times of shortfall. This could be more cost-effective than achieving the same impact with other forms of electricity storage such as batteries or hydrogen.

10.2.2 Modelling of the electricity system in the mid-2030s

There has been extensive work carried out in modelling of energy systems, the requirements for electricity storage and demand-side management (including by UK Government, UCL, Imperial College and Aalborg University). For this section, simplified modelling is presented based on daily data taken from [3] to show how DH/C systems could contribute to managing increased levels of renewable electricity generation and hence gain additional income. Data from the year 2019 was chosen as it was before the impact of COVID-19 and also has periods of extreme weather so perhaps more typical of future weather patterns.

The modelling assumes that matching of demand and supply over any 24 h period can be achieved using a combination of:

- Existing pumped storage
- Demand-side management, encouraged by time-varying electricity tariffs, including using electric vehicle charging
- Imports and exports to Ireland and the continent
- Battery storage

Hence, the modelling is all presented as average daily demands in MW.

The daily variation in GB demand in 2019 and the mix of plant supplying it is shown in figure 10.1.

Wind turbines have variable output, not just because wind speed varies but because their output is proportional to the cube of the wind speed (up to their rated output) thus amplifying the variability in the wind itself. The variation in wind strength is determined by the weather. In periods of low pressure, winds are high but will vary in location as the depressions cross the country over a 1–3 day period. At times of high pressure, winds can be low for several days or even weeks. In the UK, the inherent variability in wind generation is tempered to some extent by having a geographic distribution of wind farms from the Irish Sea to the North Sea and from Scotland to the south of England. Also, offshore wind has been found to have less variation in output than onshore wind due to the more uniform wind speeds. The variation in daily

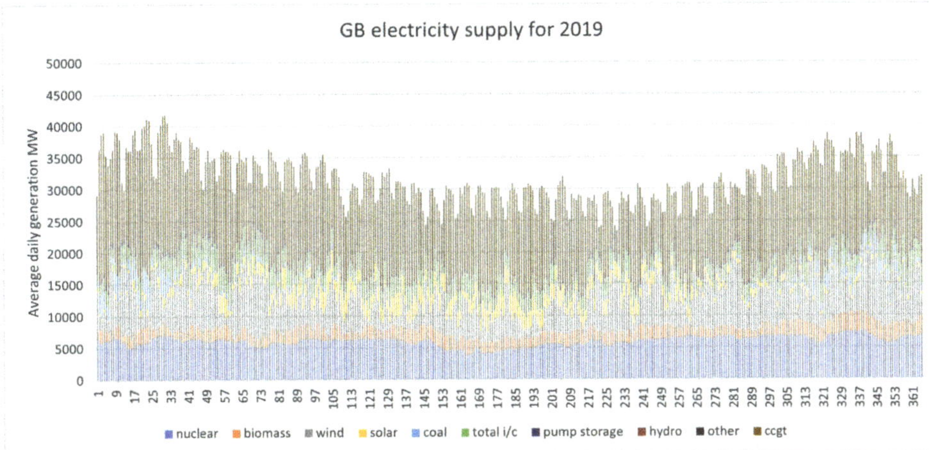

Figure 10.1. GB electricity supply as average daily generation (MW) for 365 days from 1 January 2019.

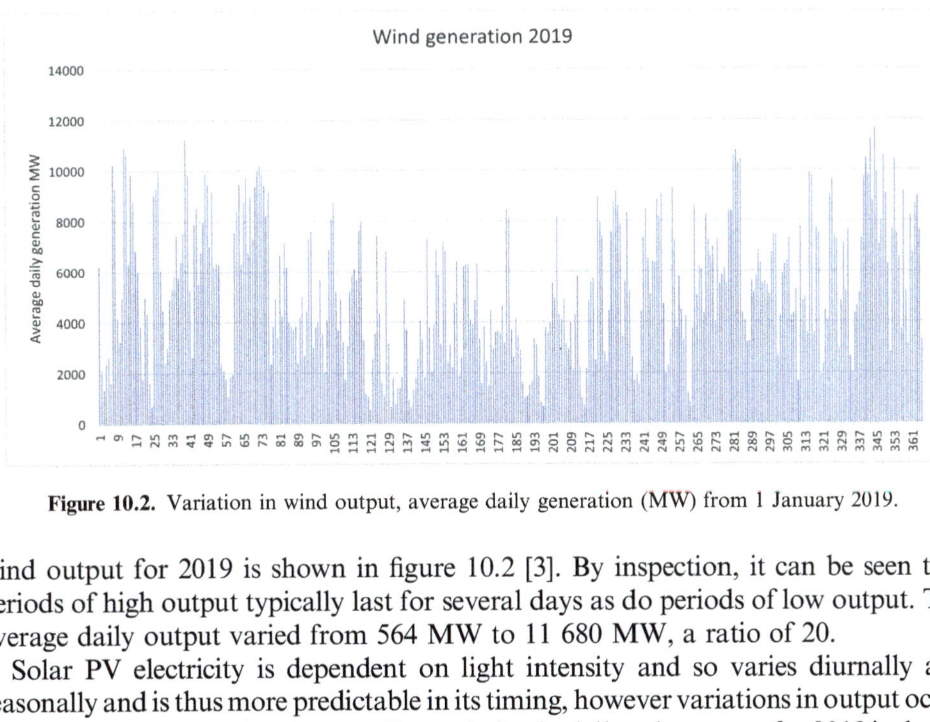

Figure 10.2. Variation in wind output, average daily generation (MW) from 1 January 2019.

wind output for 2019 is shown in figure 10.2 [3]. By inspection, it can be seen that periods of high output typically last for several days as do periods of low output. The average daily output varied from 564 MW to 11 680 MW, a ratio of 20.

Solar PV electricity is dependent on light intensity and so varies diurnally and seasonally and is thus more predictable in its timing, however variations in output occur due to the amounts of cloud cover. The variation in daily solar output for 2019 is shown in figure 10.3 [3]. The higher output of solar in summer is helpful to complement the wind output which tends to be lower in summer (note that the vertical scales on these graphs are different). The variation in solar output over the year is more significant than for wind—the average daily output varied from 85 to 3839 MW, a ratio of 45.

At present, gas CCGT plant (and a small amount of open cycle gas turbine (OCGT) plant) is used to match demand and supply and so responds both to

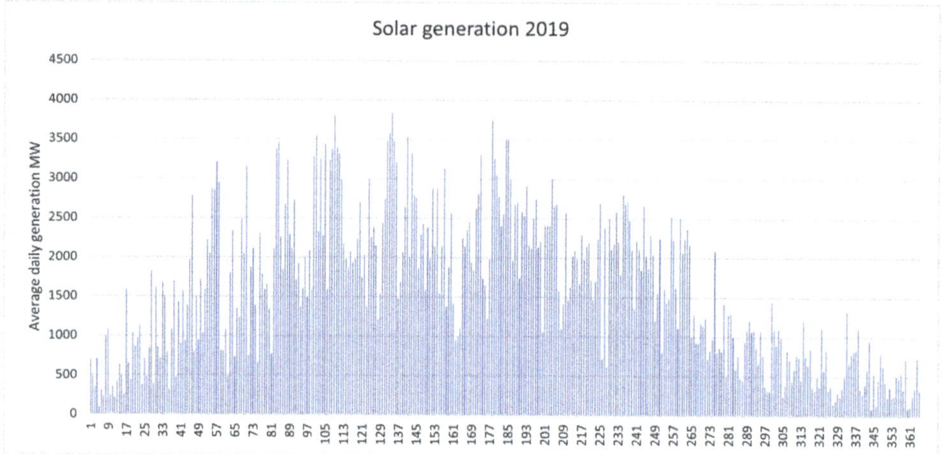

Figure 10.3. Variation in solar PV output, average daily generation (MW) from 1 January 2019.

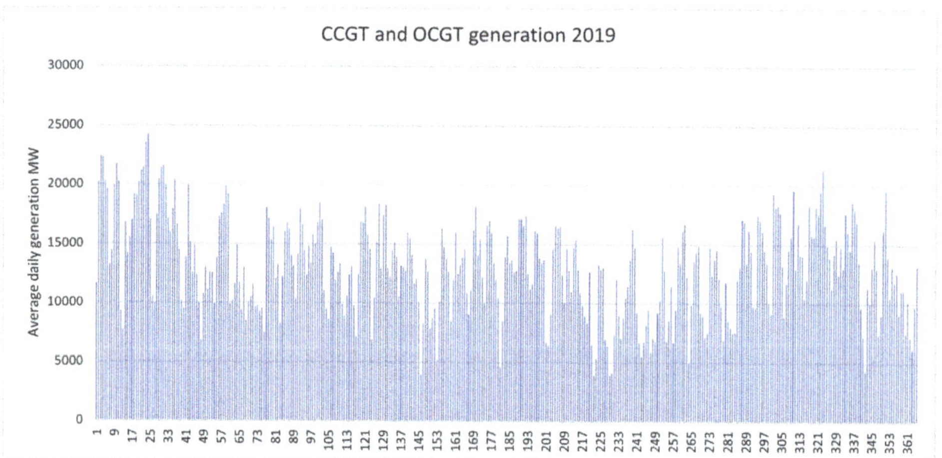

Figure 10.4. Variation in gas CCGT and OCGT output, average daily generation (MW) from 1 January 2019.

changes in demand and the variation in renewables output. As a result, the CCGT/OCGT plant operated with a daily average of between 3889 to 24 231 MW over the year, a ratio of 6, see the data for 2019 in figure 10.4 [3].

The Government strategy published in March 2023 [1] sets targets for growth in wind and solar PV. It envisages an increase in offshore wind capacity of ca. 40 GW by 2030 compared to a current total wind capacity of ca. 20 GW and a five-fold increase in solar capacity by 2035.

To illustrate a future position in the mid-2030s where wind and solar capacities have been increased a model was set up based on the 2019 data shown above. The outputs of wind and solar were increased from 2019 levels by factors of 3 and 5 in accordance with the published strategy. As there is likely to be proportionately more offshore wind than onshore wind in the future the wind energy variation would be less variable than predicted in this model and so this is a pessimistic approach with respect to the variability of wind generation.

By 2030, the remaining AGR nuclear power stations will have been shut down with only Sizewell B of the existing nuclear fleet continuing in operation. The loss of nuclear capacity is expected to be replaced with the EPR Hinkley Point C under construction and Sizewell C, a second EPR similar to Hinkley. It is also possible that small modular reactors (SMR) will be built. The model assumes a 20% increase in nuclear generation compared to 2019 which is consistent with Sizewell B (1200 MW), Hinkley Point C and Sizewell C (6400 MW) and 3 SMRs at 470 MW each, all in operation with a load factor of 80%.

Generation from biomass, other generators, pumped storage, hydro and import and export from overseas has been kept the same as for 2019. Coal use has been set to zero.

Electricity demand has been increased in the model to reflect the greater use of EVs and heat pumps. It has been assumed that there will be 6 m EV's each with an annual demand of 2.7 MWh p.a. and a uniform demand through the year. This represents a growth rate of 10% year on year from current sales of 267 000 EVs in 2022. It is also assumed that there will be 6 m additional dwellings with heat pumps, each with an annual heat demand of 9600 kWh (8000 kWh space heating, 1600 kWh water heating) and with an average CoP of 2.66. The current Government target is for 600 000 heat pumps a year, so this represents 10 years growth at this rate. In addition, new policies and capital grants are being brought forward to encourage DH with heat pumps. The electricity demand estimated for ca. 2035 is shown in figure 10.5 and the higher winter demand can be seen by comparing this with figure 10.1.

After increasing the contribution from wind, solar and nuclear and increasing the demand for EVs and heat pumps the model shows:

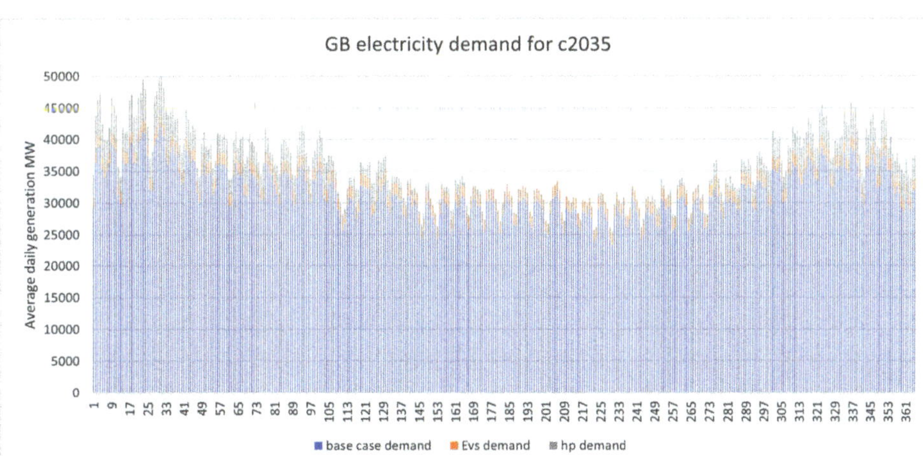

Figure 10.5. Future electricity demand, showing base case, EVs and heat pumps.

- There are periods of time when there is a potential surplus of generation and the output of generators would need to be curtailed to match demand, or directed to electricity storage.
- There are periods of time when there is a shortfall of generation and gas-fired CCGT (or some other source) will need to be brought on-line, or the shortfall could be met from electricity storage.

This is shown in figure 10.6.

For the assumptions used, the amount of surplus generation was 15% higher than the amount of shortfall so that if an electricity storage system was available with sufficient capacity and with losses of around 15%, then a zero carbon electricity supply would be achieved.

The amount of surplus electricity is 11.5% of the total electricity generated and 17.2% of the total generated from wind and solar.

The amount of shortfall electricity is 10% of the total generated. The maximum average daily power requirement for the shortfall is 32 GW and the load factor on this peaking plant would be 11.1%. If the shortfall was all met by unabated gas-fired CCGT with an emission factor of 0.46 kg kW^{-1} h^{-1} (see section 6.4), and assuming all other sources to be zero carbon, then the average electricity emission factor for this scenario would be 0.046 kg kW^{-1} h^{-1}.

The system was in surplus on 195 days and there was a shortfall on 170 days. This is illustrated in figure 10.7 by sorting the data in figure 10.6.

By inspection of figure 10.6, it can be seen that significant surplus electricity is generated over periods of several days and there are also periods when top-up is required lasting several days. There are also two more lengthy periods at the beginning of the year when there is significant shortfall and in the late summer and autumn of the year when there is significant surplus generation. In total the system

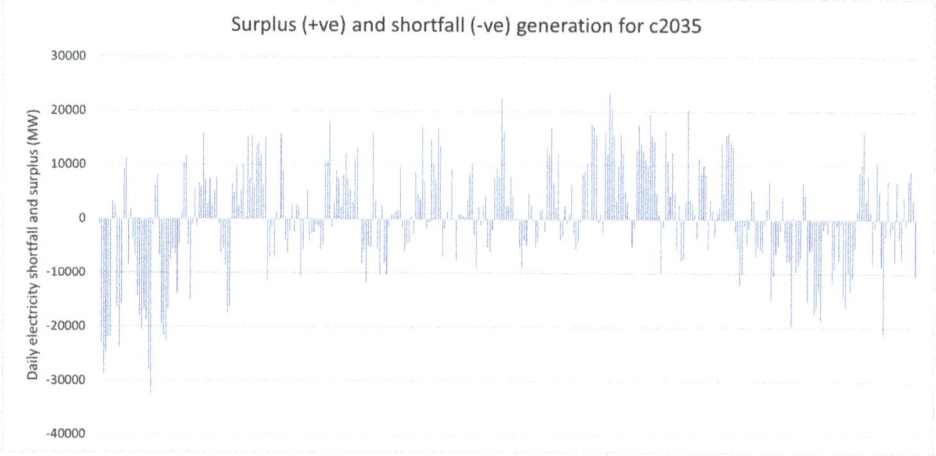

Figure 10.6. Surplus electricity available (to storage) and shortfall of electricity for a future scenario (average daily MW).

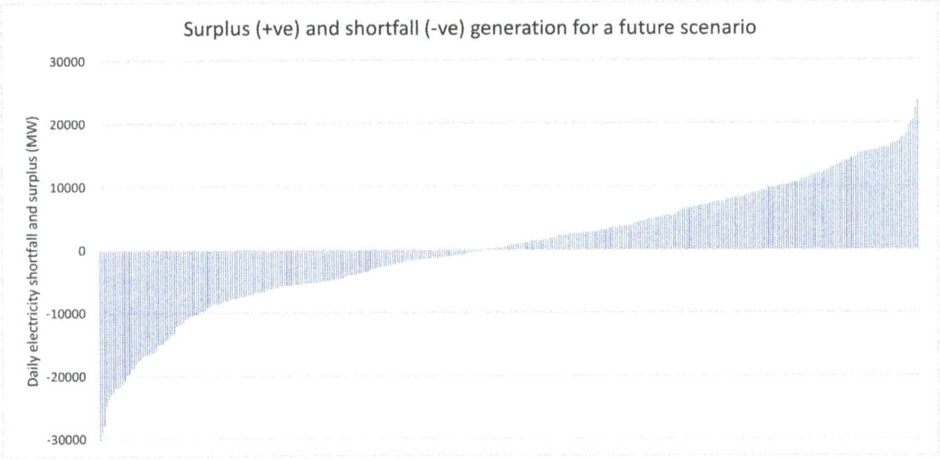

Figure 10.7. Surplus (+ve) and shortfall (−ve) sorted by value.

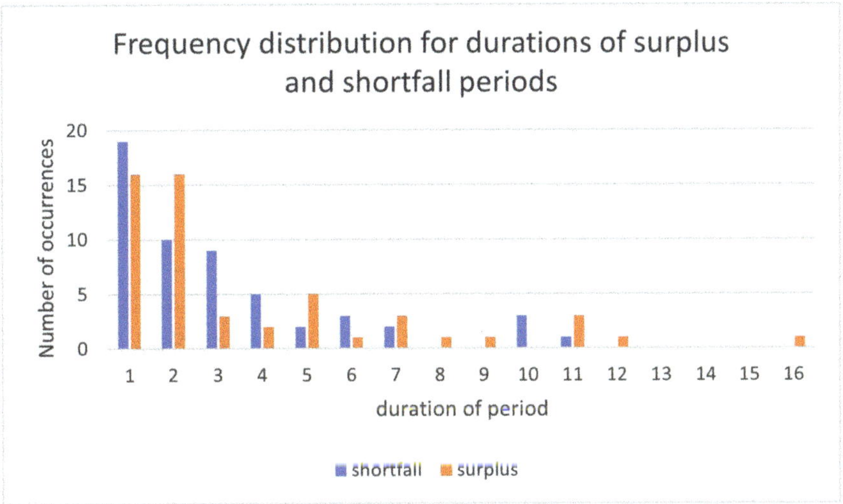

Figure 10.8. Frequency distribution for duration of periods of shortfall and surplus.

demand switches from surplus to shortfall 53 times, i.e., on average weekly, although there is considerable variation in the length of the cycles. This is shown in figure 10.8 which shows the frequency distribution for the number of consecutive days of surplus or shortfall. Most of the consecutive periods of shortfall or surplus are of less than three days duration. This is consistent with the typical variable weather in the UK. However, there are a few occurrences of 9–12 days and in one case 16 days of continuous surplus which shows that more prolonged periods of high winds or low winds can occur.

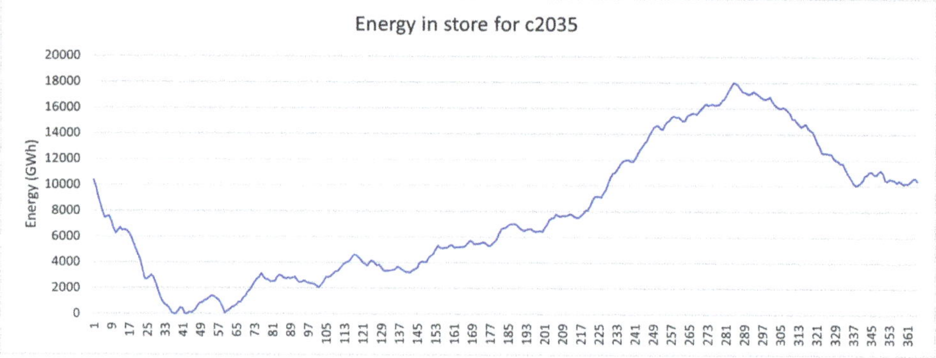

Figure 10.9. GB electricity storage required for a future scenario (assumes zero losses).

The data can be presented in another way showing how an unlimited electricity storage system would need to operate over the year to maintain an annual balance of the system as shown in figure 10.9.

In the period from Day 73 to Day 217 the storage system is both being filled and emptied in cycles with periods of days, but with a gradual increase in stored energy as there is a net surplus (this is mainly due to the higher solar contribution in summer). However, the graph also shows a seasonal imbalance. In the first month, up to Day 33, wind energy is low and the system needs to be supplied from the store. From Day 221 there is a prolonged period of surplus with the store being filled until from Day 289 there is a prolonged period of shortfall leading to the store being depleted.

It is stressed that the analysis based on 2019 shown above is an illustrative example and the variability of weather patterns mean that in any year the patterns will differ. In the future the impact of climate change is expected to result in more extreme weather patterns so even an analysis of past years will not provide an accurate forecast.

10.2.3 Sensitivity analysis

Given the large number of variables a full sensitivity analysis is not possible here, however, it is worth illustrating some impacts of varying the assumptions on demand and supply. The progress made with decarbonisation of the electricity system has been very significant especially the development of offshore wind. However, progress may not be sustained at the same rate.

In the last four years, the capacity of installed solar PV increased on average by 2% per year and the capacity of onshore wind by 3% per year; offshore wind has increased by about 1 GW p.a. since 2018 [4]. The rate of growth of offshore wind will need to increase to 4 GW per year to meet the growth targets. Growth in

Table 10.1. Assumptions for sensitivity analysis.

Assumption	Base case	Sensitivity case
Wind energy	Multiplier of 3	Multiplier of 2.5
Solar energy	Multiplier of 5	Multiplier of 3.5
Nuclear	20% increase	No increase
EVs	6 million	10 million
Heat pumps	6 million	10 million

renewable generation may be constrained by planning difficulties and by delays in connecting to the National Grid.

Large nuclear power stations take about 10 years to build so a decision to build Sizewell C will need to be taken soon to have an impact by 2035. Small modular reactors may be faster to build but they are still in development.

EVs may be taken up at a faster rate than assumed especially if there is a ban on new petrol cars. Heat pumps could also be installed more quickly than 600 000 units a year if there is a ban on using gas boilers as the replacement boiler market is about 1.7 million units a year.

Hence, as a sensitivity the model has been re-run with assumptions in table 10.1:

The results of the sensitivity analysis are shown in figures 10.10 and 10.11 for comparison with figures 10.6 and 10.7. The CCGT plant now supplies 23.8% of demand with the load factor increasing to 23.1%.

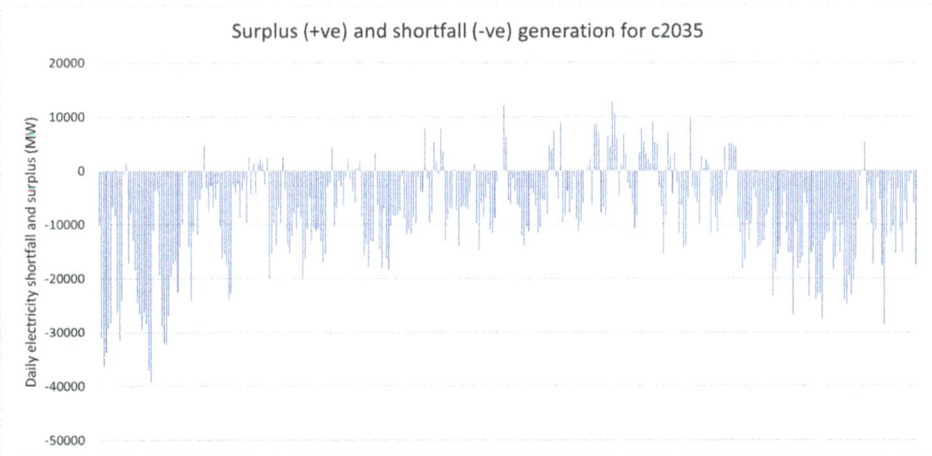

Figure 10.10. Surplus electricity available (to storage) and shortfall of electricity for a future scenario—sensitivity case (average daily MW).

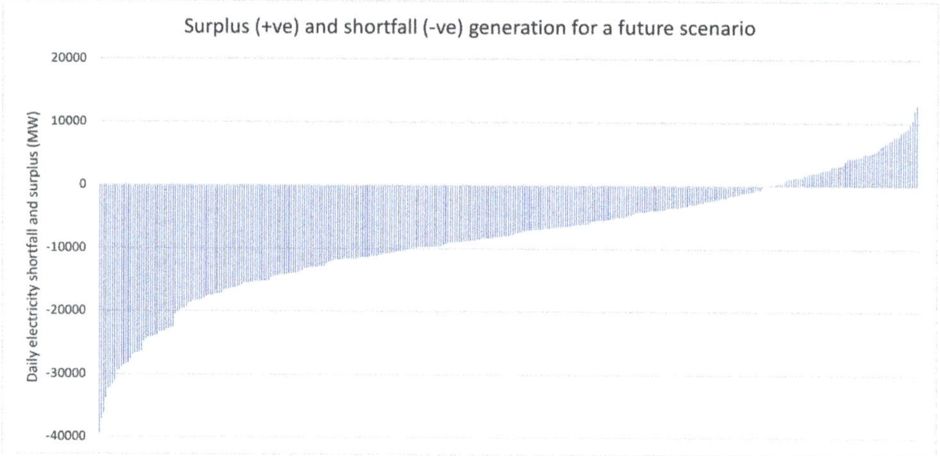

Figure 10.11. Surplus (+ve) and shortfall (−ve) sorted by value.

10.2.4 Implications for DH/C systems

Gas-fired CHP

From the base case results figure 10.7 (without long-term electricity storage) CCGT plant will be operating at the margin for 170 days p.a. with over 5000 MW of CCGT in use for 150 days p.a. this period is weighted into the winter when most DH heat demand occurs. This means that gas-fired CHP would still be able to save CO_2 at these times and would save more CO_2 than a heat pump as shown in section 6.4. The implication is that CHP should be retained to operate into the 2030s albeit for a reduced number of days in the year. For the rest of the year a heat pump would be able to produce low carbon heat.

The period of viable CHP operation would be significantly longer for the sensitivity case outlined above with 296 days when CCGT is at the margin and 233 days when CCGT output is over 5000 MW. It will be important for CHP operators and DH developers to monitor the projections and trends for demand growth, and nuclear and renewables capacity growth, carefully to enable optimal decisions to be made on retiring existing CHP or installing new capacity.

Thermal energy storage

For a heat pump based scheme lower CO_2 emissions and lower operating costs would be obtained if there was sufficient thermal storage to avoid operating in the periods when CCGT was generating as the marginal plant. The analysis shows that, in addition to diurnal variations (not modelled here), there are both short-term variations of typically less than a week and longer-term variations which can be considered interseasonal. This is supported by the report by Mott Macdonald [5] for BEIS which identified four use cases:

- Frequency response (very short response periods)
- Peak lopping (4–6 h) (diurnal)
- Long-term storage—weekly energy balancing

- Long-term storage—seasonal energy storage

To eliminate the CCGT use and reduce costs associated with curtailing renewable output will require a method of storing electricity during periods of surplus and releasing it in periods of shortfall.

The cost of storage is related to:

- The capital cost of the store
- Its efficiency, i.e., the losses in a 'round-trip' of empty to full to empty
- Its lifetime—battery storage has a lifetime of a few thousand cycles
- The number of storage cycles in a year

The capital cost of the store needs to be recovered by the cost differential between buying energy to store and selling energy to release and any other specific payments available to support storage. The number of times in a year that this net benefit is realised will determine the economic viability of the store [6]. Current battery technology appears to be close to viability in balancing the short-term diurnal variation but would be too expensive to balance power over periods of weeks as it would be used too infrequently over the year.

Two options that could provide longer-term storage are:

- Hydrogen—as natural gas has been stored successfully in geological features for many years this option is attractive especially for long-term and interseasonal storage. However, the efficiency when producing hydrogen from electricity using electrolysis and then generating electricity again from the fuel is relatively low at about 30%, (heat is released in these processes which could be used for DH).
- Large-scale TES—the ability to modify the demand profile of heat pumps (and chillers with DC) to avoid periods of shortfall and to generate heat to store during periods of surplus would have the equivalent impact of electricity storage. In terms of the use cases referred to in [5] thermal storage could play a part in: peak lopping, weekly energy balancing and, at a large-scale, longer-term seasonal energy storage. However, there are constraints as the heat pump and TES also needs to meet the heat demands of the DH customers.

If the cost of additional TES is lower than the cost of other approaches, such as hydrogen or using curtailment and gas-fired CCGT peaking plant, then this offers an opportunity for additional value to be obtained for DH/C projects. The scale of DH/C projects would be an advantage in that it will be easier to contract for demand management services with a few large DH/C operators than with large numbers of individual households.

10.2.5 Thermal store sizing

When assessing the size of a thermal store in relation to an annual heat demand a useful parameter is the thermal energy that can be stored divided by the annual heat demand expressed as MWh GWh^{-1}. Typically, thermal store sizes for DH installed in the UK were found in a survey (albeit from a very small sample) to be 15 m^3 GWh^{-1} [7]. For a 30 °C temperature difference and 90% effectiveness, this is equivalent to an energy store of 0.46 MWh GWh^{-1}.

Research carried out at UCL by Salman Siddiqui [8] showed that a much larger thermal store at about 10 MWh GWh^{-1} (i.e., a TES of 1% of annual demand) or over 20 times current practice would be cost-effective in a 2050 net zero scenario to assist in managing the variability of renewable electricity. The increase in optimal size predicted compared to existing practice is because currently TES is sized to take advantage of electricity prices and demand variations over the day rather than over several days or even weeks which will be required for storing surplus electricity (see figure 10.8).

The market mechanisms by which this benefit can be realised in the DH project economics are unclear at present. However, it would be expected that to encourage electricity storage and to reflect the need to recover capital and operating costs of the generating plant and the potential need to compensate generators for shutting down at times of surplus, electricity prices are likely to vary much more significantly than at present. The work by Siddiqui [8] shows that energy prices would be very low for 80% of the time but rise to very high levels (10–30 times the base price) at times of renewable energy shortage.

If this differential price is large enough then this will result in investment in energy storage of whatever type. For a DH system the ability to store heat over a long period, including seasonal energy storage if at a large enough scale, will mean that the production of heat by a DH heat pump could benefit from very low energy prices for much of the year. If the operation was limited only to those periods of surplus generation, with thermal storage being used in shortfall periods, then heat will be produced at a much lower cost compared to heat pumps supplying individual buildings without access to long-term storage. The UCL work [9] estimated that electricity costs to a DH heat pump would be 21% of that paid by building-based heat pumps. It would be 40% lower without thermal storage due to the supply being at HV but a further 40% reduction would be achieved through thermal storage as operating during high price periods would be avoided. This would result in a major operational cost advantage in favour of DH in the future. However, there are constraints—heat can only be supplied to storage if there is excess heat available after meeting the current heat demand of customers. The volume of heat to be stored and the times when it needs to be stored will determine whether additional heat pump capacity is required. So the capital cost of a future DH system would need to be higher for both the additional TES and potentially additional heat pump capacity.

This research [9] also shows that there are significant electricity system capital cost benefits in reducing the costs of: generation capacity, grid storage, transmission and distribution system upgrades as the TES can be used to avoid the operation of the DH heat pumps at times of peak demand.

10.2.6 Seasonal energy storage

The analysis of section 10.2.2 shows that there is a potential need for interseasonal electricity storage. In section 6.7 the potential for combining solar thermal heat generation with interseasonal heat storage, i.e., generating heat in the summer for use the following winter was introduced. This technology is appropriate for DH as it only becomes economic at large-scale. With very large volumes of water the heat losses over six months can be manageable as the surface area to volume ratio reduces with the volume stored.

It is possible that excess renewable electricity will be generated in summer (due to the growth in wind and solar PV) and a higher winter demand will occur as more space heating is transferred from gas to electricity. Interseasonal heat storage could therefore assist in managing the seasonal mismatch between renewable generation and demand as well as storage over periods of weeks as a result of variable wind generation as shown in section 10.2.2 and figure 10.9. DH heat pumps could be run in summer to store heat for use the following winter. There are three technologies available.

Pit storage systems
This technology has been developed from the waste disposal industry where landfill with contaminants needs to be stored in a long-term secure manner without leakage into surrounding soil. A shallow pit is excavated and lined with a waterproof liner. An insulated cover is installed consisting of floating polystyrene blocks. A number of such stores have been built in Denmark, e.g., the 200 000 m^3 store at Vojens which used a previously worked gravel pit (https://stateofgreen.com/en/solutions/world-largest-thermal-pit-storage-in-vojens/).

Aquifer thermal energy storage
This system relies on porous rocks to store hot water in the summer and extract it in the winter for use in DH. In addition it can be used to store cold water in the winter for use in the summer for DC. A number of examples exist especially in the Netherlands. Oslo Airport has a system which uses ground water source heat pumps linked to the aquifer for interseasonal storage for use in both heating and cooling.

Closed loop borehole storage
High temperature (40 °C–50 °C) borehole storage—in this mechanism a borehole array stores heat at relatively high temperatures—the geology of the ground is critical for this type of store as the storage needs to be in impervious bed rock rather than areas where a flow of water may remove the heat. This type of store is significantly more complex and expensive but can be more compact and allows for direct heat supply or very high CoP heat pump supply to a DH network (e.g., Drake Landing, Canada (https://dlsc.ca/index.htm#).

Geological caverns and minewater
In Vasteras, Sweden the existence of rock caverns prompted their use for oil storage in the past for energy security reasons. These caverns of 300 000 m^3 volume are now

being repurposed for interseasonal heat stores, (https://www.bbc.co.uk/news/business-65098792).

Where such natural features do not exist it may be possible to use man-made features such as coal mines, salt mines, quarries or gravel pits. The challenges are limiting the leakage of water underground and creating sufficient separation between the hot water charging and discharging the mine and the lower temperature return water to avoid mixing. A well-known example is Heerlen in the Netherlands which uses an abandoned mine in conjunction with an ambient network (see section 10.8), (https://guidetodistrictheating.eu/heerlen/).

10.3 Hydrogen

There are three types of hydrogen production:

- 'Green' hydrogen derived from electrolysis of water using electricity from renewable sources.
- 'Blue' hydrogen derived from natural gas but with the CO_2 produced captured and sequestered.
- 'Grey' hydrogen derived from natural gas but without CO_2 capture and sequestration (as is currently used for the production of hydrogen for industrial processes).

The blue hydrogen route does not result in a zero carbon fuel as the CO_2 capture process is not perfect and the upstream emissions associated with natural gas extraction, transport, liquefaction and gasification remain. The latter are not insignificant as shown in [10].

Hydrogen may in the future be blended with natural gas in the gas grid to reduce the CO_2 impact from using gas boilers. Or the existing gas grid could be repurposed area by area to deliver hydrogen with some upgrading needed due to the different characteristics of hydrogen.

The main advantage of hydrogen is that it can be stored for long periods including interseasonally without degradation and, for hydrogen produced from electrolysis of water, this could help with the issue of variability in renewable electricity generation.

In terms of electricity storage, the round-trip efficiency: electricity to green hydrogen to electricity is poor compared to batteries, about 30% compared to 85%, however, this is more than offset by the lower storage costs where interseasonal storage is required. Modelling carried out in 2020 [11] found that the use of hydrogen as a means of storing electricity would lead to a lower cost outcome.

The efficiency of conversion from electricity to heat in a boiler via green hydrogen is very poor compared to electricity to heat using a heat pump. About six times as much electricity is required if hydrogen boilers are used compared to electricity and heat pumps (see figure 10.12) [12].

The Government has the ambition to have production capacity of 1 GW of green hydrogen and 1 GW of blue hydrogen by 2025 and 10 GW capacity by 2030. The Net Zero Hydrogen Fund of £240 m has been set up [1]. Clearly the benefits of

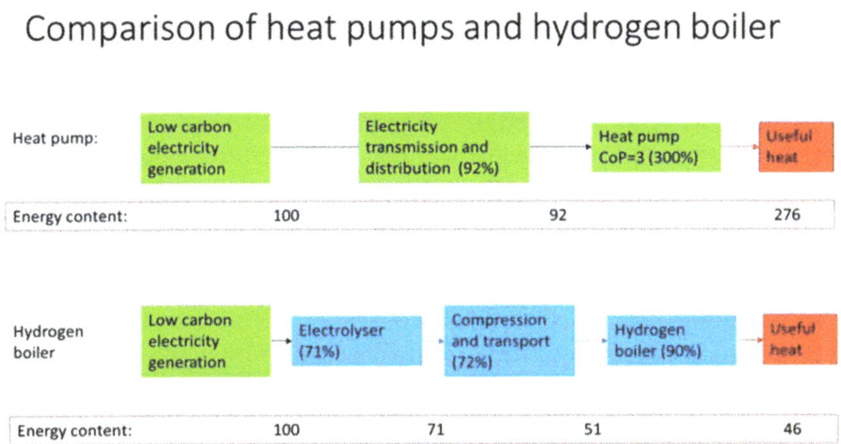

Figure 10.12. Alternative heating strategies using hydrogen or heat pumps.

reusing the gas grid and having low capital costs and a relatively small impact on householders is attractive, however, the inefficiency of producing green hydrogen will mean the fuel will be expensive.

The alternative of 'blue' hydrogen depends on the ability to capture and store the CO_2 released in its production from natural gas. This option received backing in the Government's latest energy strategy where £20 billion investment was indicated to develop CCUS from industrial processes, power plants and hydrogen production (see section 10.4).

If hydrogen was used for electricity generation at times of low renewables output then the case for using hydrogen in CHP plants for both industrial heat demand and DH would be relatively strong as the higher efficiency of CHP would yield significant economic benefits given the expected relatively high cost of electricity in these periods.

A further opportunity is the use of the significant quantity of low grade waste heat generated as a byproduct of the electrolysis process which could be recovered to supply a DH network probably involving a heat pump to raise the temperature of the waste heat [13].

The over-riding issue is that green hydrogen is likely to be a relatively expensive fuel and of limited availability so it is more likely that its use will be restricted to: replacing existing grey hydrogen demand in the chemical industry, high temperature heat in industry, HGV transport, aircraft and other applications where few suitable low carbon alternatives exist.

10.4 Carbon capture usage and storage

The Government's energy strategy Powering up Britain issued in March 2023 [1] proposes that CCUS be taken forward with up to £20 billion identified for supporting the development. This is far more than the finance available for other technologies including DH (£220 m). CCUS may be applied to power stations, industrial processes or in connection with the manufacture of 'blue hydrogen'. It is envisaged that initially four clusters will be supported where there are a number of potential carbon capture opportunities together with proximity to depleted oil and gas fields where CO_2 can be stored.

The implications for DH are that there may be long-term opportunities for heat extraction from gas CCGT, energy from waste, or biomass power stations fitted with CCUS. As described in section 6.5 the heat will be a very low carbon heat source due to the high efficiency (high z-factor). However, as always with such opportunities, the location of a major power plant using CCUS is likely to be governed by a range of factors including proximity to other CCUS plants to create clusters and the proximity to major gas and electricity infrastructure, rather than the proximity to heat demand and the supply to a DH network. Although heat can be economically transported over 10 km or more this is dependent on having a large enough demand which may need to be built up over time using smaller heat production plants.

Two clusters have been identified in Track-1:

- East Coast cluster covering Teesside and the Humber regions with CO_2 captured from a number of industries including blue hydrogen production and two gas-fired power stations with the CO_2 stored in the Endurance saline aquifer. There is potential for heat supply to Hull and Middlesborough (https://eastcoastcluster.co.uk/).
- The HyNet cluster in the north-west of England and North Wales with CO_2 captured from hydrogen production and local industries and stored in depleted gas fields in Liverpool Bay. Hydrogen will be stored in salt caverns at Northwich currently used for natural gas storage. There is potential for local hydrogen fuelled CHP plant to supply industry and DH projects in Warrington and Runcorn, (https://hynet.co.uk/).

A further eight clusters are being reviewed by the Government of which two will be selected for Track-2 development.

10.5 Small modular nuclear reactors

The Government has initiated an international competition to receive bids for funding to deliver small modular reactors as an alternative to conventional large nuclear power stations of over 1000 MW (Hinkley Point C is 3200 MW).

Rolls Royce has been proposing small modular reactors (SMR) based on the pressurised water reactor designs used for many decades to power submarines. The design has achieved Step 1 of its Generic Design Assessment and is entering into Step 2 from April 2023 with support from the Government of £210 m. The capacity of the proposed power stations each with three reactors is 470 MW.

It is likely that the first group of SMRs will be located on existing nuclear sites, with Rolls Royce proposing Sellafield, Oldbury, Wylfa and Trawsfynydd. Later sites may include ones that are closer to centres of population as they are expected to be easier to site. However, Oldbury could be suitable to supply the northern parts of Bristol.

Rolls Royce has recognised the principle of the plant being used as a DH heat source. An extract from Chapter 1 of Step 1 Generic Design Assessment [14] states:

The RR SMR has the potential to be integrated with and support:

1. *Electricity generation including baseload, load follow, frequency response and provision of other services to the power grid*
2. *Combined Heat and Power (CHP)—grid electricity plus steam off-take, for e.g., district heating*

Other designs may come forward that are at an even smaller-scale; it may be easier to find suitable sites for these which could supply DH schemes.

SMRs are likely to be delivered from the early 2030s. This would be a compatible timescale for many DH networks which may use heat pumps and/or gas-fired CHP plant to develop and grow networks over the next decade before connecting to a SMR for their long-term low-cost source of low carbon heat.

10.6 District cooling

Climate change is likely to lead to a greater need for cooling of buildings and the provision of cooling in sectors where normally this would be considered unnecessary, e.g., in housing or health care. This will create new opportunities for DC.

The case for DC is strongest where the following can be achieved:

- Where diversity of cooling demand for a group of buildings can be exploited to reduce installed chiller capacity
- Where there is a natural source of free cooling, e.g., cold river water arising from snow melt as used in Stockholm and Goteborg, Sweden.
- Where low or zero carbon refrigerants are used in contrast to HFC refrigerants in chillers in buildings.
- Where DC is combined with DH to gain an energy advantage (see section 10.7).
- Where coolth storage can be exploited to gain an advantage in electricity costs (see below).
- Where absorption chillers are viable where there is a very low carbon high temperature waste heat source.
- Where evaporative cooling towers can be used for DC but not for buildings creating an efficiency advantage—especially effective in periods of very hot dry weather. In parts of the Middle East the use of wastewater from sewage treatment plants is allowed to be used for DC as the use of fresh water is not permitted for cooling towers whereas chillers in buildings need to be air-cooled for the same reason.
- Where non-energy related benefits have been found to be significant such as release of space in the building and reduced management overheads for managing chillers and cooling towers if used; in some cases, the supply of DC can free up the architectural design, perhaps allowing a different roof design or achieving an additional occupied floor with a major financial benefit.

The main challenges for DC are:

- Limited utilisation of high capital cost assets over the year when meeting non-process cooling demand.
- DC networks are more expensive than DH networks as the temperature difference in the network is limited to around 10 °C resulting in larger pipes for the same energy flow.
- The differential in capital and operating costs between centralised chillers and building level chillers may be insufficient to justify the cost of the DC network.

Coolth storage

Chilled water storage can be used to lower operating costs by using low-cost electricity at times of low electricity demand, e.g., at night and in the future at times of high wind/solar generation. Chilled water storage is preferable to ice storage because there is a cost penalty in operating the chillers at the lower temperatures needed for freezing. Phase-change materials may be used for smaller schemes where space is a premium.

Very large chilled water stores have been used for DC in the Middle East and the Far East. In addition to steel, concrete has also been used to construct the tanks which is possible as the temperatures are low (see also section 6.14).

10.7 Linking of DH and DC

Where a group of buildings have both cooling and heating demands in the same season both DH and DC networks can be installed independently to supply these as in figure 10.13.

Efficiency gains can be obtained by linking heating and cooling networks together. In principle there are five possible approaches as illustrated in figures 10.14–10.18:

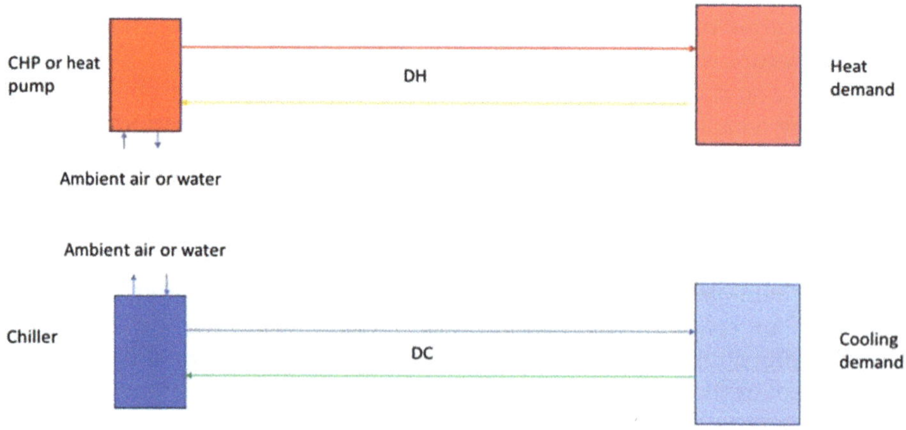

Figure 10.13. Separate DH and DC.

(A) Using heat rejected from the DC chillers as part of the heat source for the DH heat pumps
(B) Using the heat pump heat source circuit to pre-cool the DC return water
(C) Using a heat driven chiller (absorption or adsorption) at the Energy Centre to supply DC taking the DH as the heat source
(D) Using heat driven chillers in each building taking the DH as the heat source
(E) Using an ambient temperature network with heat pumps and chillers/heat exchangers in each building using the ambient network to take heat as a source for a heat pump or to reject heat from the chiller/heat exchanger

All of these circuits assume that there are days in the year when both heating and cooling demands exist, for example space cooling and domestic hot water heating. It is also often the case in shoulder months that cooling demands are seen in the afternoon and heating in the night and early morning. In this situation diurnal thermal storage may be used to enable the linkage between the two systems. However, methods A and E can also be used in conjunction with interseasonal storage including aquifer thermal storage (see section 10.2.5).

Some of these methods (A and E) have been implemented in large buildings where both heating and cooling demands may occur simultaneously. However, in many groups of buildings there may be a demand for cooling in one building, say an office block and a demand for domestic hot water heating in an adjacent residential block. It is only by using a form of DH/C that the surplus heat in one building can be utilised in another.

(A) Using heat rejected from the DC chillers as part of the heat source for the DH heat pumps

Figure 10.14 shows a system where the heat rejected from the chiller circuit is used as the heat source for the heat pump. This arrangement is well known for large buildings where internal spaces need cooling and perimeter spaces need heating. An example is the New British Library, London.

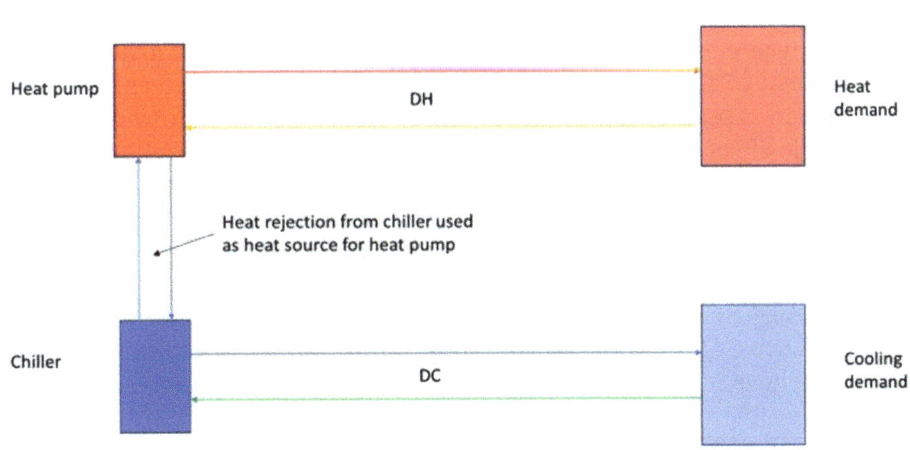

Figure 10.14. Chiller reject heat as source for heat pump.

(B) Using the heat pump heat source circuit to pre-cool the DC return water

Figure 10.15 exploits the low temperature in the return to the heat source from the heat pump as a direct contribution to the cooling circuit, thus reducing the load on the chillers. This is easiest to establish from say a river or wastewater heat pump system where the water may be discharged at a temperature low enough to pre-cool the return water in the DC circuit. In some cases, the river water itself may be cool enough to provide free cooling without the use of a heat pump. The water returned from the heat exchanger is warmer which improves the heat pump CoP. An example is the Hammerby plant in the Stockholm DH/C system which uses wastewater as its heat source. The Stockholm DC system also draws cold water from depth in the adjacent fjord.

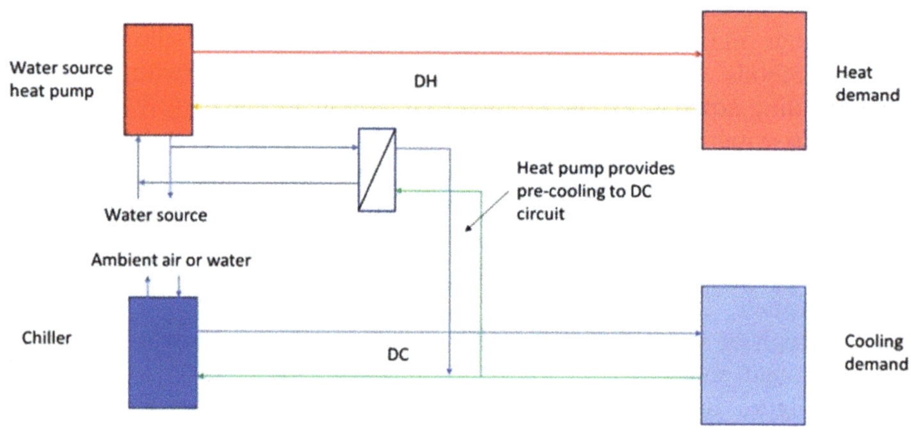

Figure 10.15. Heat pump heat source circuit pre-cooling DC circuit.

(C) Using a heat driven chiller (absorption or adsorption) at the energy centre to supply DC taking the DH as the heat source

Figure 10.16 uses heat driven chillers (absorption or adsorption) to produce cooling at the energy centre. This is viable provided the CO_2 content of the heat is low enough to compete with a conventional electric chiller. Absorption chillers have a relatively low CoP of around 0.7 for single-effect and 1.1 for double effect and are more cost-effective for high temperature heat supply. As a result, they have been used successfully with CHP plant. In the UK, with current electricity emission factors the use of absorption chillers with gas-engine CHP is unlikely to give a carbon saving. However, where the heat is derived from steam extraction or flue gas heat recovery at an energy from waste plant with a high *z*-factor of say 7-10 the effective CoP for a single-effect absorption chiller would be about 5-7 and a double effect chiller about 8-11. This could be better than the best achievable with conventional chillers and also avoids many of the issues around refrigerants. An example is the Queen Elizabeth Olympic Park

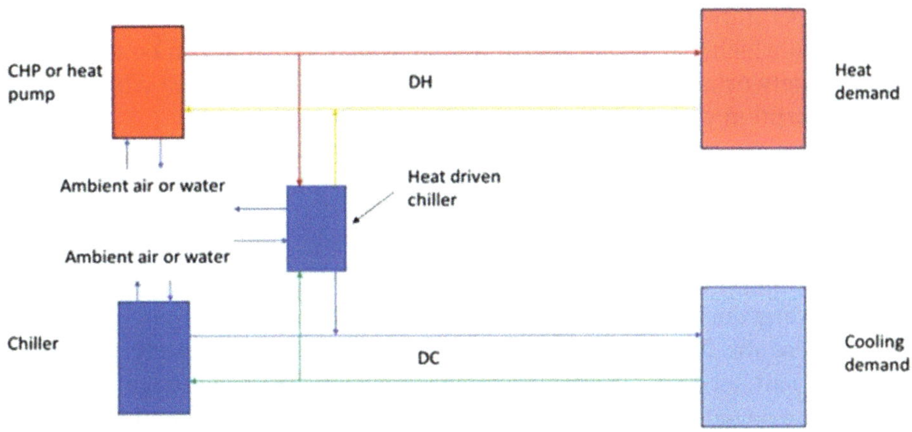

Figure 10.16. Heat driven chillers for DC.

which has a double effect absorption chiller in the Kings Yard Energy Centre and the Goteborg DC system using heat from an energy from waste plant to supply centralised absorption chillers.

(D) Using heat driven chillers in each building taking the DH as the heat source

Figure 10.17 uses heat driven chillers at each of the buildings. This arrangement has the major advantage of removing the need for a DC circuit and its associated capex. As for figure 10.11 the CO_2 content of the heat supplied needs to be very low for there to be a CO_2 benefit. The building will need to have sufficient space to install not just the chiller but the heat rejection equipment which will be around twice the capacity needed for conventional plant. In addition, the temperature at which the DH network needs to operate will typically need to be higher to supply the heat driven chillers than would otherwise be the case when there would be benefits in lowering the flow temperature in summer. The DH-side

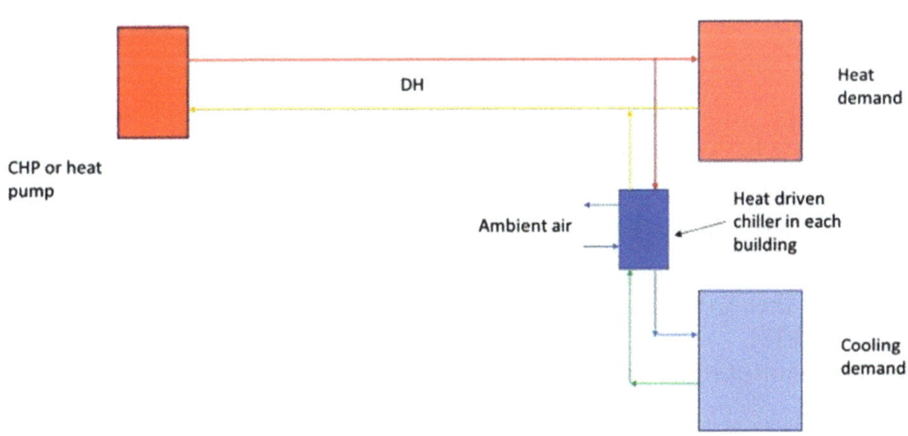

Figure 10.17. Distributed heat driven chillers.

temperature difference achieved with absorption chillers is also relatively small which may constrain the amount of cooling that can be supplied from the network. An example is the Seoul DH system in Korea where the use of absorption chillers by customers has been encouraged.

(E) Using an ambient temperature network with heat pumps and chillers in each building using the ambient network to take heat as a source for a heat pump or to reject heat from the chiller.

Figure 10.18 uses a single ambient temperature network reducing capital costs and this network is used both as a heat source for heat pumps in each building and a heat rejection circuit for the cooling system in each building. Where the heating and cooling demand are approximately the same the ambient network temperature would typically be in the range 15 °C–25 °C. However, at the supply point there could be a need to add heat or reject heat to maintain the overall energy balance. This circuit has the potential to be a cost-effective way of using heat rejected from one building to provide heat to adjacent buildings. It is, therefore, discussed further in section 10.8. An example is the South Bank University Balanced Energy Network. (https://www.cibsejournal.com/technical/intranets-for-heat-introducing-ben-networks/).

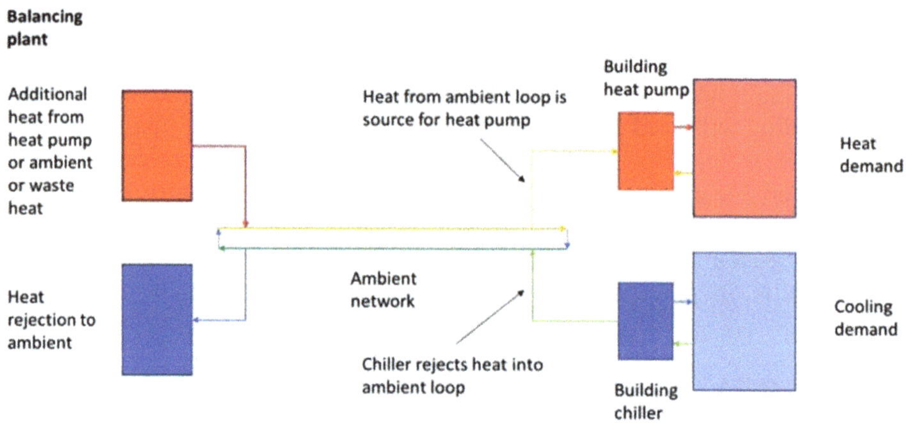

Figure 10.18. Ambient temperature network.

10.8 Ambient networks

The principle of ambient networks is to distribute water at temperatures near to ambient and then to use heat pumps in each building to provide the heat required (see figure 10.18).

Ambient temperature networks offer a number of advantages:

- As discussed in section 10.7, where there is a need for cooling and heating at the same time the network can function as a heat sink as well as a heat source improving efficiencies for both heating and cooling. i.e., the heat rejected from a building's cooling system can be reused by other buildings as a heat source. With conventional DH/C this can be achieved but only with two networks, DH and DC.

- Often in commercial office building applications there is a need for heating in the morning and cooling in the afternoon. So a heat pump used in the morning will reduce the temperature in the ambient network so that there is a lower temperature heat rejection circuit available for use by chillers or heat exchangers in the afternoon.
- Heat losses from the network are much lower than conventional DH as they operate closer to the ground temperature, this means that insulation thicknesses can be reduced or avoided altogether.
- The heat pumps can be designed and operated to deliver heat at the temperature required for each building (e.g., a newer building may not need such a high flow temperature) whereas a centralised heat pump system will need to produce heat at a temperature to suit the building on the network that needs the highest temperature even if others could use a lower temperature, this improves heat pump efficiency.
- As some of the required temperature rise has been achieved by the network it is easier for a heat pump to produce higher temperatures from a single stage process.
- The heat pumps can be installed internal to the building so there is less impact on the surroundings or roof space compared to ASHPs.
- A DH system may need to operate at a minimum of 65 °C to deliver DHW from the central source whereas a heat pump within a building can operate at a lower temperature to suit the space heating demands, except for short periods when DHW needs to be heated.
- It is possible that an ambient network could be developed primarily for heating and then linked to buildings which can reject heat to it later so that a better energy balance is obtained.
- The customer is normally charged a fixed cost for the provision of the ambient temperature circuit and is responsible for paying the electricity costs for the heat pump as part of their normal electricity supply. There is no need for a separate metering and billing service as for a DH system.
- Ambient networks are well suited to ATES as the temperatures are suitable and heat pumps and boreholes can be distributed.

However, there are also disadvantages:

- Building owners will have to provide space for the heat pumps and in many cases associated domestic hot water storage; this could be a significant cost and disruption disadvantage especially in dwellings.
- The design temperature difference between flow and return is less and so pipe sizes will be larger (the impact on trench costs is partly offset by having thinner or no pipe insulation).

- It will be more difficult to introduce thermal energy storage (TES). The ΔT is smaller on the network so storage will need to be downstream of the heat pumps within the buildings where space is constrained. It is possible that storage using phase-change materials could be incorporated as this does not require a large ΔT.
- Multiple heat pump installations within the buildings may be more expensive and complex than having centralised heat pumps especially for existing buildings and for dwellings there is the added cost of a hot water cylinder.
- The additional electricity demand for the heat pump may require an upgrade of the electricity supply to the building.
- Maintenance of multiple small heat pumps may be more expensive than maintaining centralised heat pumps and may put more responsibility on the customer, depending on the form of contract.
- Network costs will normally be covered by a fixed charge and customer's electricity bills will cover the cost of running the heat pump. To market the concept to customers it will be necessary to show that the higher CoP for the heat pump compared to say an ASHP will more than cover the fixed charge for the ambient network supply. This is in contrast to a DH system where the heat is sold on a p kWh^{-1} basis, and can be more readily compared with other heating systems.
- The range of refrigerants available for use in buildings may be more restricted than for a centralised heat pump which may have implications for efficiency and environmental impact.
- Although it will be straightforward to reject heat to the circuit there may be disadvantages in injecting heat if the heat source is at a temperature significantly higher than that of the ambient network. For example, if a datacentre is rejecting heat at 35 °C it would be more efficient to use a heat pump to raise the temperature to 70 °C for distribution by a conventional DH network than to dilute it with ambient network temperatures of 15 °C–25 °C.

The advantage of using the network as a heat source and a heat sink is of particular interest in the context of rising cooling demands especially for large residential or mixed use developments where there will be a demand for domestic hot water in summer. The benefits of ambient loops tend to diminish where cooling (and/or heating) demands are not distributed geographically within the proposed network.

In general, it is unlikely that there will be a perfect balance between heating and cooling and although there is some storage in the network and additional storage can be added there is usually a need to inject additional heat into the network or to reject heat from the network. The additional heat can be provided by say an air source heat pump and heat rejection can be by means of a dry air-cooler. An ATES can provide a method to balance demands over the year.

It can be seen from the above that ambient temperature networks are quite different to the district heating concept which has evolved from first to fourth generation. As a result, the term fifth generation could be misleading in the sense that it is not necessarily superior to fourth generation. The 5G terminology has been questioned in a paper by Professor Lund *et al* [15] which describes it as a complementary technology rather than a chronological successor.

10.9 Zoning policy

For DH, DC and ambient networks the economic case will be strongest when all buildings in a given area are connected. This is especially true if the opportunities to reduce network costs described in section 10.10 are to be implemented where clearly all houses in a street will need to be connected.

In Denmark, in response to the oil price rises in the 1970s heat planning was introduced where it was decided on the basis of cost studies which areas of the city would have DH and which would be heated with oil or natural gas. However, in Finland, when the alternative to DH was using expensive oil boilers, very high levels of market penetration were achieved without the need for mandating of customers to connect.

In the UK, at present, building owners cannot be compelled to connect except as a result of planning conditions. For example, at the Queen Elizabeth Olympic Park a planning condition for all buildings being constructed in the LLDC area is that they are to connect to the DH and DC networks. Also, until recently, DH has had to compete with gas-fired boilers with relatively cheap fuel costs.

However, the Government has indicated that they will include in the Energy Security Bill currently being debated, the requirement for DH zones to be established in towns and cities—see the policy statement on the heat network zoning framework [16].

The zones will be defined using standardised software which will identify where DH would be more cost-effective than the alternative means of achieving low carbon heating. This will not limit DH to these areas but will avoid the situation where some buildings install their own heat pumps (and in sufficient numbers to require an area-wide electricity network upgrade), others retain their gas boilers and some connect to a DH network. In combination, this scenario would clearly be less cost-effective for all parties and inhibit growth in DH.

Within these zones it is expected to be made compulsory for certain buildings to connect to the heat network. The buildings that may be mandated to connect are: all new buildings, large public sector and commercial buildings, and multi-occupancy residential buildings which already have communal heating, or are undergoing major refurbishment. There will also be an exemption process. At present the existing housing market is not to be mandated to connect through zoning, however, confidence in DH technology in this market segment will be raised if all major buildings in the area are connected, so mandating may not be necessary. However, some design solutions that would bring down costs significantly depend on all houses in a street being connected (see section 10.10).

The regulations may also include a requirement on heat sources within the zone to connect to the network where feasible. This is important as there is currently little incentive for organisations with a suitable heat source to supply a DH scheme as there would be little commercial gain for them.

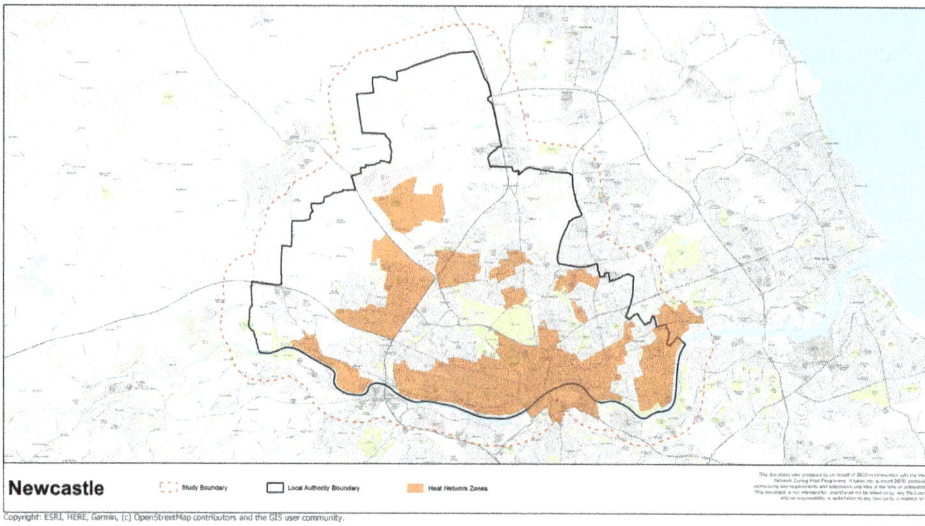

Figure 10.19. Sample results from zoning work for Newcastle. Reproduced with permission from Newcastle City Council.

A national emissions limit for heat networks, with scope for lower local emissions limits, is likely to be part of the zoning regulations.

The Government carried out a consultation on zoning in 2021 and has published its response [17]. A pilot project is currently underway (from April 2022) with some initial outputs shown in figure 10.19 with a further consultation on the detail of the regulations planned for 2023. The Heat Networks (Scotland) Act 2021 [18] has also provided the powers for introducing regulations on zoning of heat networks in Scotland.

10.10 Opportunities to reduce DH capital costs

10.10.1 Overview of opportunities

Both solar PV and offshore wind have seen significant falls in capital costs over the last 10 years. There is little evidence of similar reductions in costs for DH/C. This is partly because every project appears to be unique and requires both bespoke designs and significant development costs.

Some parts of a DH/C system have become more standardised. Pre-insulated pipework is manufactured at scale and the materials cost is now a relatively small part of the total network cost. Greater use of plastic carrier pipes should lead to faster construction with skills in pipe jointing transferable from other industries. Building interfaces are now pre-assembled in a factory as HIUs or thermal substations and their components are mass produced. Boilers, pumps, valves, pipework, controls and switchgear in energy centres are all standard products produced competitively.

Some areas where cost reductions may be available in the future are:

- Large heat pumps where greater volumes may be produced and where there is scope for technical development in the light of experience.
- The cost of excavating trenches and back-filling could benefit from greater use of trenchless technology, e.g., directional drilling, and better information on existing buried services.
- Development costs could be reduced with a more streamlined approach to obtaining funding from Government and other bodies. The standardisation of draft contracts issued by the Government through the HNIP is an example.

Future market share for DH/C will depend on how its costs compare with those of individual building heat pumps. In any comparison of options, the costs associated with upgrading the local electricity network to enable widespread use of building level heat pumps should be considered. If zoning is introduced, the zones will be defined as the areas where DH is shown to be the lowest cost solution to deliver low carbon heating.

A project led by the Energy Technologies Institute [19] identified eight 'route maps' which could assist in reducing the costs of DH. These are:

- Knowledge management—establishing a DH Knowledge Centre to share learning and disseminate results from the other route maps.
- Low flow rate design—ensuring that design is optimised to avoid oversizing and maximise ΔT in heating circuits.
- Radical routes for pipework in housing (see section 10.10.2).
- Trenchless technologies (e.g., directional drilling).
- Improved front end design and planning—obtaining better data on existing buried services in advance of construction to save time on site.
- Sharing of civil engineering costs with other utilities through greater co-ordination and better planning.
- Greater use of direct connection and reuse of existing hot water cylinders.
- Optimising the design of HIUs through simplification and design for manufacture.

Overall, a cost saving of 32% was estimated from the above. The ETI's modelling also showed that the impact of such a cost reduction would be to increase the market share of heat networks from 31% to 43% by 2050.

10.10.2 Radical routes

Conventional designs use pre-insulated pipe buried in the street with branches to individual houses.

In more suburban areas houses are further apart and are set further back from the road so the network costs will increase simply as a matter of the geometry. The increases in pipe length are offset to some extent by the greater heat demand from larger houses whether semi-detached or detached houses.

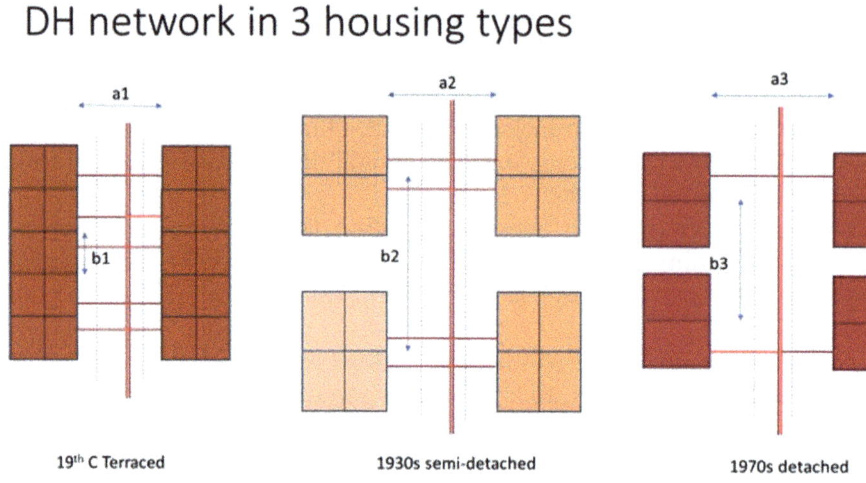

Figure 10.20. Typical geometries for three sample house types.

Table 10.2. Length of heat main per MWh of heat demand for three housing types.

Housing type	Length of street main per house (m)	Length of branch main per house (m)	Total length per house (m)	Annual heat demand per house (kWh)	Total length per MWh per house (m MWh^{-1})
19th C terraced	$b_1/2 = 2.4$	$a_1/2 = 6$	8.4	7000	1.2
1930s semi-detached	$b_2/4 = 4.5$	$a_2/2 = 10$	14.5	11 000	1.3
1970s detached	$b_3/2 = 5.5$	$a_3/2 = 12$	17.5	15 000	1.2

At the scale of local street mains and branches the cost of the pipe is mainly determined by its length with the diameter of the pipe having less impact on cost. So a useful parameter relating to cost is 'length of pipe installed per MWh of heat supplied' (the reciprocal of linear heat density). This can be calculated for different housing typologies.

The layouts and dimensions in figure 10.20 and table 10.2 uses dimensions taken from streets in Cardiff and shows the difference in this cost parameter moving from terraced housing to detached. It was found that the 1970s detached houses were built closer together than the earlier semi-detached houses and even though the total length is twice that of the terraced houses this is compensated by the higher heat demand, resulting in similar lengths per MWh supplied.

Table 10.2 shows that the branch pipes to individual houses become more significant in more suburban areas. So any methods which result in shorter lengths of branch pipe per house will be of value in extending the reach of DH.

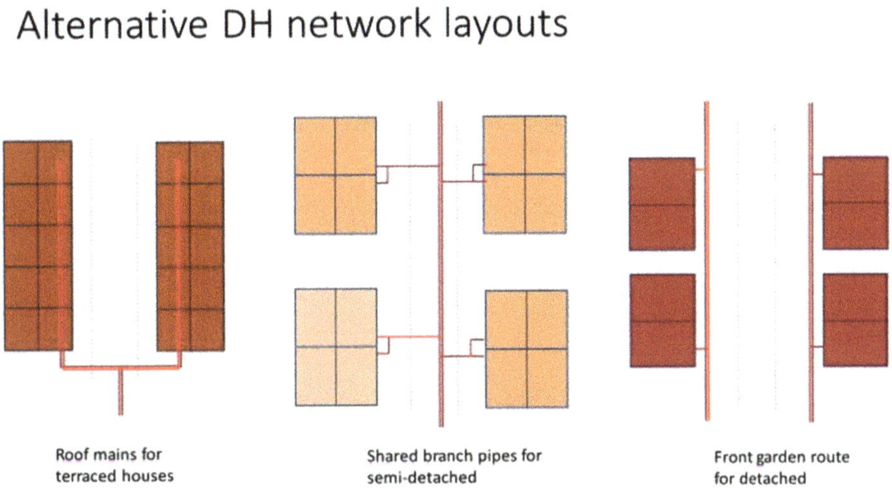

Figure 10.21. Alternative routes for DH pipes in three sample house types.

In the ETI work [19], three potential opportunities to significantly lower cost were identified (see figure 10.21):

1. Roof mains

   The mains would be run within the loft spaces of terraced houses, eliminating the costs of trenching, back-filling and reinstatement in the street and resulting in very short internal branches. This method could also be used for semi-detached houses where these are spaced closely together with gable roofs so that a short pipe bridge can be formed between the houses.
2. Shared branch pipes

   Applicable to semi-detached layouts where one branch from the street main would supply two semi-detached houses, approximately halving the branch length
3. Front garden routes

   This route could be used for detached houses where these are set back from the road and where there are driveways rather than extensive landscaping. Two mains along the street would be installed close to the front of the houses. The street mains would be double in length but the branches would be very short, resulting in an overall saving.

Table 10.3 shows that for these revised pipe layouts each dwelling type has very similar lengths per MWh heat sold. The reduction in total length for each alternative design by comparison with table 10.2 indicates a potential for cost savings of around 30%. Further savings would be obtained for the roof space route as the civils costs of trenching and reinstatement are avoided. The implication is that with innovative designs lower density housing could be connected economically, increasing the potential market for DH.

These designs assume that all houses in the street will be connected to DH and that suitable legal provisions can be put in place for the shared pipes to pass through private land or within other properties, similar to that which exists for other utility services.

Table 10.3. Length of heat main per MWh of heat demand for three housing types—alternative routes.

Housing type	Length of street main per house (m)	Length of branch main per house (m)	Total length per house (m)	Annual heat demand per house (MWh)	Total length per MWh per house (m MWh^{-1})
19th C terraced roof mains	$b_1 = 4.8$	1	5.8	7	0.83
1930s semi-detached shared branches	$b_2/4 = 4.5$	$a_2/4 = 5$	9.5	11	0.86
1970s detached front gardens	$b_3 = 11$	2	13	15	0.87

The district heating wall

An alternative to roof mains for terraces is to locate the pipes externally on the front wall above the doorway. The visual impact could be minimised by suitable cladding of the pipes. Similar advantages to the roof mains would be realised but with potentially less disruption to the occupant and easier access for any maintenance work. It also avoids potential issues where the roof space has been converted to an attic room or maybe in the futue.

The visual impact would be avoided entirely if this method was combined with external wall insulation. The wall insulation would also provide the necessary insulation for the pipes. This opportunity was studied by Frost *et al* [20] and is illustrated in figure 10.22 and is referred to as the 'district heating wall'. The

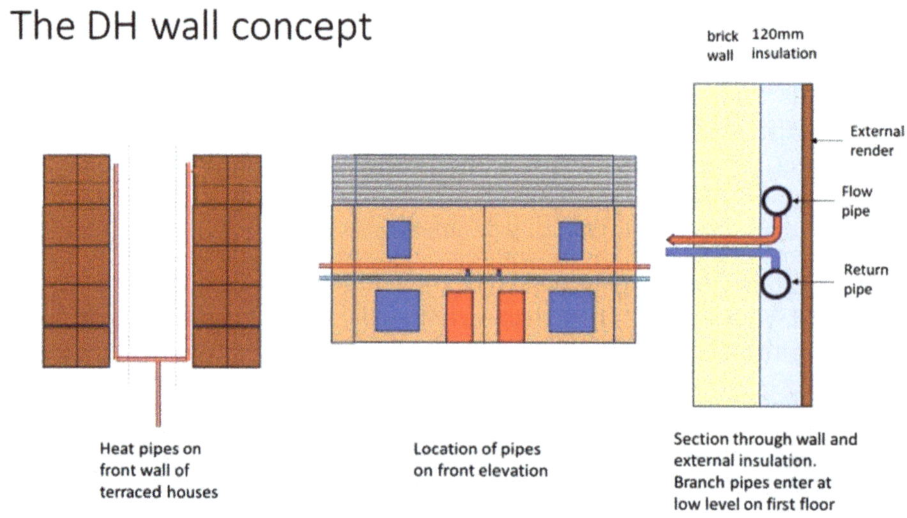

Figure 10.22. District heating wall concept.

important synergy of this solution is that it would lead to both lower heat demands as well as much lower capital costs for DH. The insulation would enable lower temperatures or lower flow rates to be used with the existing radiators retained, improving the overall cost-effectiveness of the DH network. This solution could be applied to the 2.5 million solid wall terraced houses which are likely to be in high heat density areas close to city centres.

Clearly this approach does require all properties to connect to DH and for the external wall insulation to be fitted to all properties in the street.

Bibliography

[1] Powering up Britain *HMG March 2023* (https://www.gov.uk/government/publications/powering-up-britain)

[2] Elliott D 2016 *Balancing Green Power* (Bristol: IOP Publishing)

[3] DataMaximum Ltd 2023 *Daily Outputs for Generators Downloaded for 2019* Supply and Processing of Datasets (https://datamaximum.co.uk/)

[4] BEIS 2022 *Digest of UK Energy Statistics* 2022 chapter 6 and table 6.2 (https://gov.uk/government/statistics/digest-of-uk-energy-statistics-dukes-2022)

[5] Macdonald M 2018 Storage Cost and Technical Assumptions for Electricity Storage Technologies (https://www.gov.uk/government/publications/storage-cost-and-technical-assumptions-for-electricity-storage-technologies)

[6] Gates B 2021 *How to Avoid a Climate Disaster* (Allen Lane) pp 76–7

[7] DECC 2015 *Assessment of the Costs, Performance, and Characteristics of UK Heat Networks Final Report*

[8] Siddiqui S 2022 Modelling district heating in a renewable electricity system *Dissertation for the Degree of Doctor of Philosophy, Energy Institute* (University College London)

[9] John M and Mark B 2023 The impact of Thermal Energy Storage (TES) on the economics of heat networks using large heat pumps and zero-emission electricity *UCL Energy Institute Research Report* (UCL)

[10] Mark B and Tizziano Gallo C 2021 Heating with steam methane reformed hydrogen *UCL Energy Institute Research Report* (UCL)

[11] BEIS 2020 Modelling 2050 electricity system analysis (https://gov.uk/government/publications/modelling-2050-electricity-system-analysis)

[12] Cebon D *A Comparison with Heat Pumps (Part 1)—Hydrogen Science Coalition Hydrogen for Heating?* (https://h2sciencecoalition.com/blog/hydrogen-for-heating-a-comparison-with-heat-pumps-part-1/)

[13] Ramboll 2021 Synergy study heat recovery from hydrogen production *Ramboll for UK Danish Embassy* (https://storbritannien.um.dk/en/the-trade-council/projects/district-heating/heat-recovery-from-hydrogen-production)

[14] Royce R 2023 *E3S Case, Chapter 1—Introduction E3S Case Chapter 1: Introduction* (https://gda.rolls-royce-smr.com/assets/documents/documents/rr-smr-e3s-case-chapter-1---introduction-issue-1-gda-website-version.pdf)

[15] Lund H, Østergaard P A, Nielsen T, Werner S, Thorsen J, Gudmundsson O, Arabkoohsar A and Mathiesen B V 2021 Perspectives on fourth and fifth generation district heating *Energy* **227** 120520

[16] HMG 2023b Energy Bill Policy Statement Heat Network Zoning, DESNZ, March 2023 *Energy Security Bill: Policy Statements and Draft Regulations.* (https://www.gov.uk/government/publications/energy-security-bill-policy-statements-and-draft-regulations)

[17] Department of Business June 2022 *Energy and Industrial Strategy, Heat Networks Proposals for Heat Network Zoning Government Response* HMG (https://gov.uk/government/consultations/proposals-for-heat-network-zoning)

[18] Scottish Government 2021 Heat networks (Scotland) Act 2021 *Scottish Government, 2021 Heat Networks—Renewable and Low Carbon* https://www.gov.scot/policies/renewable-and-low-carbon-energy/heat-networks/

[19] ETI 2018 District heat networks in the UK *Potential Barriers and Opportunities* (Energy Technologies Institute)

[20] Frost 2012 The 'District Heating Wall': a synergistic approach to achieve affordable carbon emission reductions in old terraced houses *Low Carbon Econ.* **3** 115–29

IOP Publishing

An Introduction to District Heating and Cooling
Low carbon energy for buildings
Paul Woods, MA MSc CEng FIMechE FEI FCIBSE

Chapter 11

Summary—the case for DH/C as part of the national energy strategy

11.1 Summary of the chapters

Part I—Introduction to district heating and cooling (DH/C)

Chapter 1

The development of DH in Europe was given fresh impetus in the 1970s as a result of the two world oil price rises. New sources of heat had to be found urgently. For countries dependent on oil for heating and without indigenous fuels the energy efficiency and security of supply advantages of using waste heat from power stations to heat cities was very attractive. When extracting heat from power stations, the fuel used per unit of heat delivered was found to be about a fifth of that needed when using boilers (section 1.2.1).

In the UK, we were more fortunate as we had access to North Sea gas which has met most of our heating needs for almost 50 years. We had little reason to develop DH when fuel was plentiful and cheap. As a result, our energy system continues to be inefficient. Despite the rise in wind and solar energy, thermal power stations (gas, coal, biomass, waste and nuclear) still account for 63% of our electricity supply and the heat lost in power generation is equivalent to the heat required by all of the domestic buildings in the UK (section 1.2.1).

Many UK DH projects have been developed, especially over the last 10 years, with valuable experience gained—even though the share of the market is still very small compared to other countries such as Scandinavia, Russia and China (section 1.3). Various studies have assessed the economic potential of DH in the UK to be between about 20% and 40% or 5–10 m dwellings (section 1.3.3), compared to around 3% now.

doi:10.1088/978-0-7503-5286-4ch11

Chapter 2

The technology of DH/C is well established with products that have been progressively developed over the last 40 years. Pre-insulated pipe systems offer good efficiency and a long life whether the carrier pipes are steel or plastic (section 2.1.4). A wide range of heat sources can be used to supply DH, including: extraction of heat from energy from waste plants, energy efficient devices such as CHP and heat pumps and heat recovery systems from industry and data centres as well as renewable energy sources including solar (section 2.1.5).

A DH scheme often uses a number of different types of heat source as well as thermal energy storage (TES). This enables significant cost reductions to be achieved, compared to individual building systems where this complexity would not be justified or practical. For example, with DH, the primary low carbon plant such as a heat pump does not need to be sized to meet the peak demand as peak boilers can supplement the heat pump in colder periods. Even with the primary plant sized at 50% of the peak demand it will still supply about 85% of the annual demand (section 2.1.6). This arrangement provides a practical economic solution albeit with some increase in emissions. However, if the better economics results in many more schemes being built the overall emissions will be much lower.

Part II—The engineering of DH and DC systems

Chapter 3

The heating systems within the buildings to be connected to DH are a critical component of the overall system. Whilst most existing buildings can be connected to DH some modifications to the existing heating systems are needed (section 3.5). These are principally:

- To reduce flows through heat emitters to lower return temperatures so that a circuit designed for 80 °C/60 °C can operate at 80 °C/50 °C requiring a 33% reduction in flow.
- To change the control systems to use two-port control valves instead of three-port diverting valves.
- To remove any fixed bypasses.
- To resize domestic hot water systems to operate with lower flow temperatures.

These changes will result in lower DH return temperatures and allow DH flow temperatures to be reduced in summer. This will result in lower heat losses and more efficient heat production especially from heat pumps.

The type of domestic hot water heating used will determine the DH minimum operating temperature. For instantaneous hot water heating a flow temperature of 55 °C is sufficient to deliver water at 50 °C but for hot water storage systems a temperature of 65 °C is needed to ensure that the stored water can be kept at 60 °C (section 3.3.3).

Chapter 4

Interface equipment is required to connect a building to the DH network. A direct connection means that the DH water flows into the building's heating system whereas an indirect connection involves a physical separation by means of a heat exchanger. The heat exchangers can be sized so that the approach temperature difference is 3 °C or less.

Decisions need to be made about where hydraulic separation is best placed, depending on the type of building, and the height of the building (section 4.2).

Interface equipment for dwellings is pre-packaged into a heat interface unit and a system of testing has been introduced which has encouraged innovation and progressive improvement (section 4.3.2).

Chapter 5

The importance of careful route selection based on pre-construction surveys and research is highlighted (section 5.1). The heat network is an energy transport system and it can be optimised by sizing the pipes to minimise the total cost of capital, heat losses and pumping energy (section 5.5). Heat losses within residential blocks need to be minimised by reducing the lengths of pipes installed using more risers and siting the HIUs near to the risers (section 5.4.2). The various limit states in EN13941 are discussed together with the fundamentals of stress analysis for steel pipes (section 5.7). The impact of operating temperatures on the lifetime of plastic carrier pipe systems is shown with reference to EN15362 (section 5.6.2).

Chapter 6

Gas-fired CHP and heat pumps are both energy efficient systems. Methods of comparing gas-fired CHP and heat pumps are described. Where gas-fired CCGT plant is operating as the marginal plant on the grid (as at present) CHP offers better efficiency and lower emissions than heat pumps (section 6.4). A wide range of other heat sources are available for DH including heat recovery from industry, energy from waste plants and data centres. Renewable energy sources may become more widely used in the future including solar thermal plants and deep geothermal sources (sections 6.5, 6.6 and 6.7).

The benefits of using multiple heat sources are explored and it is shown that an energy centre with both heat pumps and CHP would deliver the lowest emissions; as the amount of renewable energy increases on the system the CHP would operate less and the heat pump more (section 6.9).

Chapter 7

The selection of operating temperatures is an important design decision which influences the design of the network and energy centre. Whilst each scheme will have different optimal conditions some broad conclusions have been established. For an existing customer a network flow temperature of 85 °C should be sufficient and a 55 °C return temperature should be achievable, potentially lower in summer (section 7.2).

An illustrative calculation showed that reducing the return temperature to 45 °C would lower total network costs by 11%. Reducing the flow temperature to 75 °C will increase costs by 11% as a larger pipe size is needed and as a result heat losses are almost the same even with the lower temperature (section 7.3).

The main reason for using lower temperatures is to reduce the cost of heat production from certain types of heat source particularly heat pumps. Using a variable flow temperature which only reaches 85 °C at times of peak demand and reduces to 70 °C for most of the year is likely to be optimal for heat pump systems (section 7.5). If peaking boilers are used then if these are installed in series they could be used to raise the flow temperature with the heat pumps continuing to operate at 70 °C throughout the year.

Chapter 8

It will be increasingly important to optimise the operation of the central plant as the impact of greater variable renewable energy will result in wide variations in prices and marginal CO_2 emission factors. Greater use of thermal storage and energy centres with both CHP and heat pumps as well as cooling systems add to the potential complexity and the opportunities for optimisation (section 8.1). Water treatment will remain key to a long system life and non-chemical approaches are available (section 8.2). Heat metering has evolved from simply recording energy use to providing a monitoring tool to identify poor performance in HIUs and heating systems with the potential for two-way communication (section 8.3).

Part III—Future opportunities and the case for DH/C

Chapter 9

DH/C is able to deliver improved security of supply, lower cost and lower environmental impact. The various emission factors used in calculating CO_2 savings were reviewed and all were found to have limitations. An hour-by-hour marginal calculation is recommended rather than calculations based on a single emission factor for electricity. DH can offer lower CO_2 emissions than building heat pumps where elevated temperature heat sources are available for DH heat pumps or where steam extraction from thermal power stations is available. The wider choice of refrigerants for DH heat pumps in DH and chillers in DC will also offer an environmental advantage over individual systems (section 9.2).

It is likely that achieving lower energy costs will be the strongest argument for DH/C, especially for dwellings where the cost of installing a heat pump is high as modifications to heating systems will be needed (section 9.3).

There are several reasons why DH could be attractive to customers in addition to lower costs. There will be less disruption and fewer changes needed to existing heating systems, there will be greater trust in the company providing the DH as it will be delivering for the whole city area and the transaction costs will be low. Future reliability and maintenance will be guaranteed through the contract for the long-term supply of heat with no unexpected repair bills (section 9.5).

Chapter 10

Since the first feasibility studies were carried out in the 1970s it has taken a long time for DH/C to be seen as an important strategic low carbon solution for heating but we have now reached the point where the Government's 2023 energy strategy has described it as 'vital for making net zero a reality'. Future opportunities for DH/C can be identified from this energy strategy. There is a target to increase offshore wind capacity by 40 GW and a fivefold increase in solar energy. The addition of these variable energy sources will mean the issue of balancing and electricity storage will be ever more important. There will be times when renewable energy is in surplus when electricity prices will be low and CO_2 emissions low and other times when there is a shortfall and gas-fired plant will need to be operated, unless there is a method of storing electricity over several days or weeks. Of the options available for storage, green hydrogen is promising as it can be stored long-term in bulk. Battery storage is most likely to be used for diurnal variations. There is a role for DH with TES to provide a way of altering the electricity demand pattern sufficiently over several days for it to contribute to the management of the renewable electricity. As this will have a system wide benefit, this should in the future provide additional funding for DH schemes which will not be available for individual heat pumps in buildings where TES will be much more limited if at all (section 10.2).

The simplified modelling presented also shows that there is a role for gas-fired CHP as well as heat pumps into the 2030s as gas-fired CCGT remains on the margin for about half of the year, and it will be important to monitor trends to see how the national electricity supply systems develop. A sensitivity analysis with faster growth rates for EVs and heat pumps and a slower growth rate for wind, nuclear and solar shows that CHP would be viable for most of the year in the 2030s (section 10.2).

Other opportunities for DH are to obtain heat from thermal power stations whether small nuclear reactors or other power stations fitted with CCUS or from hydrogen production. The challenge will be whether such large-scale facilities are located close enough to large heat loads (sections 10.3, 10.4 and 10.5).

There are a number of ways in which meeting cooling demands as well as heating in an optimal manner can achieve additional savings including by the use of ambient temperature networks (sections 10.7 and 10.8).

The new zoning policy planned to be introduced by the Government is a further opportunity that will create greater certainty in the market and lead to faster growth of DH (section 10.9).

Other opportunities to reduce costs for DH include rethinking the way in which we supply houses with more radical routes including roof mains, or routing along the front elevations of terraced houses possibly within a layer of external wall insulation (section 10.10.2)

11.2 The case for DH/C

11.2.1 Background

The case for district heating can be based on delivering against the three elements of the energy trilemma:

- Security of supply
- Cost of energy
- Environmental impacts

Over the years the emphasis on each of these elements has changed but all remain important today. In the 1980s the priority was on saving energy for security of supply concerns in the light of the oil price shocks of the 1970s. In the 1990s energy prices were relatively low but since 2000 energy prices have been of concern and fuel poverty has become a dominant issue. Recently, energy prices have been extremely high due to the conflict in Ukraine leading to large Government subsidies. This has shown that our continuing reliance on fossil fuels and imports of energy carries a major risk to the economy. Since the Rio summit in 1992 climate change has been of increasing concern and DH has been able to deliver measurable CO_2 savings through CHP and in the future this will be achieved with heat pumps using low global warming refrigerants.

DH/C scores highly on all three parts of the trilemma and can also provide a good customer experience, be developed at speed and retain flexibility into the uncertain future.

11.2.2 Security of supply

In 2004 we started importing gas as our North Sea gas reserves declined. The recent conflict in Ukraine has shown Europe that dependence on outside sources for its energy has risks. Even though the UK has been shielded from a direct threat to supplies due to imports from Norway and LNG from Qatar and the USA, we have been impacted by the price rises as more of our gas is traded as LNG.

The need to move away from gas as our own reserves are depleted and the need to decarbonise has led to a strategy of electrification of heating originally set out in the first UK heat strategy published by the Government in 2013. This was the first time that a roadmap was set out to develop heat networks and individual heat pumps. From 2010 progress in developing renewable electricity sources was rapid, especially with solar PV where 12 GW was added between 2010 and 2017 and offshore wind where 10 GW has been added.

The higher efficiency available through heat pumps (compared to direct electric heating) will make best use of our electricity in meeting heat demands, whether through DH or individual building heat pumps. The continued use of gas-fired CHP in DH will be justified whenever gas-fired power generation is being displaced and CHP will reduce gas use and minimise imports of LNG improving our energy security position.

In the future, heat networks will increasingly use renewable electricity to power heat pumps as well as making use of heat recovery from power plants (especially energy from waste plants), industrial processes and commercial processes such as data centres. DH is the technology that will allow the most efficient use of indigenous energy from a wide range of sources including waste heat. As we use more resources from our own land and seas our dependence on others will reduce.

11.2.3 Cost of heating and cooling

For much of the period from 1980 to 2000 the UK benefited not just from a secure energy supply but also from relatively low costs of energy. However, there were always concerns about low-income households who lived in energy inefficient homes and the Warm Homes and Energy Conservation Act 2000 included a requirement to produce a fuel poverty strategy. This led to a continued emphasis in keeping fuel bills low and the provision of grants to assist in energy efficiency improvements. Other measures were introduced including the winter fuel payments for pensioners. Domestic fuel and electricity had the lowest possible VAT rate applied. Whilst these policies protected the fuel poor it did little to stimulate investment in energy efficiency or district heating. Other countries took a different route where they imposed both high rates of VAT and a carbon tax on domestic energy and protected their industries from high energy costs. This enabled the cost of DH and building insulation to be more easily justified.

The recent price rises as a result of the sudden return to growth after the Covid pandemic and the conflict in Ukraine has shown again that high energy prices are politically unacceptable and Government subsidies have been used to limit price rises. Recognition that long-term prices will remain at a high level has already led to new investments in energy efficiency and solar PV. There will also be renewed interest by customers in alternative thermal energy sources and new investments in district heating and cooling.

The basis economics of district heating is that a relatively high capital investment is needed to deliver heat with low operating costs and low emissions. In the past, the difference in operating cost between DH supply and the cost of heating from gas had to be sufficient to justify the capital investment. So when gas prices are low it is harder to make the case.

Although DH is still assessed against the benchmark of heating by gas boilers, increasingly it is recognised that a comparison with heat pumps in individual buildings is a more valid comparison. Against this comparator DH performs more strongly especially in the residential market. A DH scheme is still capital intensive but then so are individual building heat pumps and in addition, there are likely to be additional costs in changing building heating systems and in upgrading local electrical infrastructure.

Therefore, the strongest part of the case for DH in the longer term will be its ability to deliver heat at a lower cost than individual building systems. These lower costs arise because of higher efficiency and economies of scale in the central plant both of which need to justify the additional costs of the thermal networks.

The capital cost of a DH system is likely to be lower than individual building heat pumps due to:

- Economies of scale leading to lower heat pump costs.
- Supply of heat at a temperature compatible with existing building heating systems reducing the costs of conversion to heat pumps.
- Heat is supplied at a rate that allows for the instantaneous production of hot water avoiding the need for a new hot water cylinder in dwellings.
- Use of hybrid systems with peak boilers reducing heat pump capacity.
- Minimal space requirements in the home and garden.
- No need to upgrade local electricity infrastructure, a single new supply to an energy centre will be much lower cost than multiple building upgrades.
- The bulk of the investment has a very long life.
- Access to a wider range of finance with low interest rates.

The operating cost for DH is likely to be lower than individual building heat pumps due to:

- Where DH uses the extraction of heat from power generation plants, e.g., energy from waste, this has an equivalent CoP of 7–10 compared to 2.5 to 3.5 for heat pumps so in these cases energy costs are at least halved.
- Where heat pumps are used, electricity is bought in bulk at HV at lower prices.
- Where CHP is used, this will generate at times of high electricity price providing valuable revenue.
- Thermal storage allows more flexible operation which can avoid purchasing electricity at times of high prices and maximise income from CHP.
- Use of elevated temperature heat sources, e.g., data centres, industrial waste heat, rivers will improve efficiency compared to air source heat pumps in individual buildings.
- Lower maintenance costs as minimal equipment is within buildings and the complex plant is all in a central location benefiting from economies of scale.

11.2.4 Environmental benefits and climate change

In a future where the electricity grid is fully decarbonised and where both individual building heat pump systems and DH heat pump systems are using low carbon electricity there will be little to choose between the two systems in terms of CO_2 emissions. The higher efficiency of DH heat pumps is offset by network losses and the need to operate at higher flow temperatures. However, in the transition period until the grid is decarbonised, a DH scheme that uses both gas-fired CHP and heat pumps together with thermal storage will have lower emissions (see section 6.9).

The current strategy to decarbonise the heat sector is to convert buildings from using gas-fired boilers to electric heating on the assumption that electricity is being decarbonised through a combination of wind, solar and nuclear sources. The use of heat pumps for heating is a more efficient way of using low carbon electricity than direct electric heating and these can be used for both individual buildings and to supply DH. It is likely that DH will predominate in higher density areas in towns and cities and individual heat pumps will be used in lower density and rural areas.

Whereas air source heat pumps are likely to predominate in individual buildings, DH systems can be supplied from a much wider range of low carbon and renewable sources offering higher efficiency and so put less pressure on the electricity supply system. For example, extraction of heat from the energy from waste and biomass CHP plants, solar thermal, deep geothermal, heat recovery from industry and data centres and the use of water sources at elevated temperatures will all result in higher efficiency and lower demand for electricity and lower CO_2 emissions when compared to ASHPs.

When combined with significant thermal storage the use of DH can enable the greater use of variable renewable generation and limit the use of fossil fuel peaking plant (see section 10.2).

In addition, these efficiency improvements will lead to less demand for electricity and thus fewer embodied CO_2 emissions in the construction of wind and solar plants, transmission systems and electricity storage.

Overall, the case for DH to deliver CO_2 savings is arguably stronger than the use of individual heat pumps in buildings.

Other environmental impacts are also important. Large-scale heat pumps and chillers for DH/C systems can use refrigerants with very low global warming potential including the natural refrigerants of ammonia and CO_2.

11.2.5 The customer experience

Often analysts compare energy strategies from a national standpoint without taking account of the views of customers. But without customers being willing to change to new technology it will be very difficult to move away from gas for heating. Regulations may be introduced but they will need to be widely supported if they are to succeed. Previous regulations which led to rapid change are the move away from leaded petrol and the requirement to use condensing boilers.

In addition to low energy costs and environmental credentials, heat customers will be looking for a heat supply that is:

- Safe
- Reliable
- Quiet
- Easy to install without undue disruption
- Easy to purchase and arrange the installation (low transaction costs)
- Low maintenance costs and certainty on these (no surprises)
- Rapid response from a maintenance team if there is a problem
- Has a long life, preferably with a long-term guarantee

When compared to individual building heat pumps DH scores very highly on all of these.

When the counterfactual assumption was continuation of gas boilers then clearly the implementation of DH faced a number of barriers, e.g., lack of understanding of the technology, risks of low market penetration and opposition to the disruption caused by the streetworks to install heat mains. However, when the counterfactual is installing individual heat pumps the comparison is more in favour of DH as there is likely to be significant disruption within the building to change heating systems to operate at lower temperatures, in dwellings to install hot water cylinders and to find space for air source heat pumps that will not cause noise disturbance to neighbours. In addition, there may be some wider disruption as a result of upgrading of electricity supplies. The barrier of unfamiliar technology is likely to apply to both individual heat pumps and DH.

Customers need help to convert to a new technology and a lack of trust in the supply chain may mean a slow response. DH would be provided by a company operating across the town or city often with the backing of the local authority (LA) if not the LA having a major share. The offer will be a unique one so transaction costs will be low and contracts will be standardised and backed by regulations for customer protection.

The DH company will provide a maintenance service in the local area for a large number of properties so will employ permanent locally based staff able to provide a rapid response. The more complex plant is installed in the energy centre so maintenance of customer equipment is much reduced. Meter reading and billing is largely an automated process. The DH company will be able to guarantee a long-term supply of heat and provide fixed maintenance costs only varied by indexation. Reliability from DH is high due to multiple heat sources used.

11.2.6 Speed of change

We are running out of time to limit the worst impacts of climate change, in fact some commentators say the tipping point has passed already. So speed of change to reach a low carbon energy system is critical.

If zoning regulations for DH are introduced with mandating of connections, then there is the potential for DH to be implemented much more rapidly than individual heat pumps where progress will depend on the decisions of many individual building owners and their response to the various incentives and market mechanisms that may be introduced. They may also opt to wait until their boiler needs replacing.

DH could overcome lack of trust in new technology which holds back change as DH would be backed by local or national government on a large scale.

11.2.7 Flexibility

We have already mentioned that an energy centre containing CHP, heat pumps and thermal storage has considerable flexibility in responding to the marginal plant on the grid, using CHP to displace CCGT when this is on the margin and using heat pumps when renewables are in surplus. Thermal storage will enable this to happen on an hour-by-hour basis through the day as well as over longer timescales of several days, contributing to balancing renewables by changing demand patterns.

In addition, a DH/C system will be able to respond to advances in technology rapidly as it is only the central plant that needs changing to introduce a new technology. For individual buildings a change would involve convincing multiple stakeholders. Chapter 10 identified a number of future technologies such as CCUS, SMRs and hydrogen. All of these have the potential to be connected to DH and their energy efficiency enhanced as a result.

In addition, once a heat network is established it is easier to introduce heat recovery from industrial sources whereas basing an entire network on one source would be commercially risky. A DH/C network could be seen as a marketplace where thermal energy with a wide range of sources and customer demands can be traded. If an ambient network is developed for heating then cooling can be incorporated over time and offers flexibility as heating and cooling demands change in the future.

Table 11.1 summarises the case for DH/C.

Table 11.1. Comparisons of DH/C with individual building systems.

Issue	Individual building heat pumps	DH/C
Security of supply	• Limited action possible to improve efficiency • ASHP most commonly used which will have poor efficiency in cold weather leading to continued use of gas-fired peak power generation	• Higher efficiency and wider range of heat sources means UK renewable sources will go further, reducing need for imports of fuel • Where CHP used, this also reduces imported gas as it is more efficient
Cost of heating	• Electricity costs based on retail tariffs • Limited scope for demand side management to get lower electricity costs • Capital cost and maintenance costs higher • Difficult to raise finance	• Higher efficiency plant • Thermal storage and electricity bought in bulk leads to lower electricity costs • Economies of scale for central plant reduces capex and opex • Finance available at low interest rates
Environmental impact	• ASHP mostly used which are least efficient type • Refrigerant leakage • Noise pollution possible	• Lower CO_2 emissions from more efficient sources available from: energy from waste, data centres, deep geothermal, solar thermal, industrial waste heat, minewater • Use of low global warming refrigerants
Customer experience	• Lack of trust in supply chain	• DH company trusted by whole town or city area

(Continued)

Table 11.1. (*Continued*)

Issue	Individual building heat pumps	DH/C
	• Disruption as may need new radiators and hot water cylinder, loss of amenity space inside and outside • No long-term guarantees • Uncertainty on maintenance • Dependent on a single machine which may break down	• Minimal impact on building heating system • Long-term heat supply contract • Local maintenance team always available • No unexpected maintenance costs • High level of reliability from multiple heat sources • Prices controlled by regulation
Speed of change	• Limited by individual decisions which may take time given shift to unfamiliar technology • Change likely to happen only when boiler needs replacing	• Can be introduced over a wide area quickly where supported by zoning regulations and local authority planning policy
Flexibility	• Single heat source, probably ASHP, unlikely to change this in the future	• Multiple heat sources and thermal storage • Can switch to a different technology in the future without changing network or customer interface

11.3 Knowledge sharing

If DH/C is to fully deliver its potential there will need to be many new entrants to the industry whether in design, construction, operation, project management, legal or financial roles. We need to build capacity in the industry at all levels if we are to expand as rapidly as is now necessary. In the past, when experience was seen as a hard won commercial advantage, we perhaps did not share knowledge as freely as we could have, which may have held the industry back. With the support of CIBSE, a number of guidance documents related to DH/C have now been published including the CIBSE/ADE Code of Practice for heat networks. The HIU testing regime managed by BESA is another example where the process in itself has led to greater understanding and technical improvements. The Heat Networks Technical Assurance Scheme will also promote quality and lead to improvements over time. The ETI called for a knowledge centre to be established for the industry that would monitor and assess recently completed projects, carry out research work to improve performance and lower costs and disseminate the results rapidly—this is urgently needed.

If this book has contributed to the knowledge sharing process by raising awareness, helping new entrants make a contribution more quickly and promoting best practice, I will be very pleased.

www.ingramcontent.com/pod-product-compliance
Ingram Content Group UK Ltd.
Pitfield, Milton Keynes, MK11 3LW, UK
UKHW051858100926
8556IPUK00005B/676